Henry Evers

Steam and the Locomotive Engine

Henry Evers

Steam and the Locomotive Engine

ISBN/EAN: 9783743435711

Printed in Europe, USA, Canada, Australia, Japan

Cover: Foto ©berggeist007 / pixelio.de

More available books at **www.hansebooks.com**

Putnam's Elementary Science Series.

STEAM

AND THE

LOCOMOTIVE ENGINE.

BY

HENRY EVERS, LL.D.,

PROFESSOR OF MATHEMATICS AND APPLIED SCIENCE, CHARLES SCIENCE SCHOOL,
PLYMOUTH, AUTHOR OF "NAVIGATION," "NAUTICAL ASTRONOMY," ETC.

NEW YORK:

G. P. PUTNAM'S SONS,

FOURTH AVENUE AND TWENTY-THIRD STREET.

1873.

PREFACE.

THIS book is intended to give beginners an insight into the simple theory and arrangement of the Locomotive Engine. It has been the chief aim of the writer to make the subject as easy, practical, and perspicuous as possible, by omitting whatever is dry in theory, if not absolutely requisite, or likely to puzzle and confuse.

The author has keenly felt that a work of this kind was greatly wanted, and it has been his endeavour to supplement the design of the Publishers, in seeking to occupy a foremost place in supplying, in the best manner possible, the want so widely felt, so as to bring the Series of Text-Books, of which this volume forms one, within easy reach of every Student of Science in the Kingdom.

H. E.

January, 1873.

CONTENTS.

CHAPTER I.

HEAT.

Definition—Expansion of Bodies by Heat—The Liquid and Gaseous States of Matter—Co-efficient of Expansion—Energy of Atomic Forces—Practical Illustrations—Temperature of Bodies, and Instruments for Measuring Temperature—Thermometers—Comparison of Thermometers—Graduation of Thermometers—Pyrometers—Capacity for Heat of Bodies—Calorimeter—Conversion of Heat into Work and Work into Heat—Consumption of Heat in Liquefaction and Vaporisation—Convection of Heat—Methods by which a Large Mass of Water may become Heated—Conduction of Heat—Good and Bad Conductors—Experimental Illustrations, - - - - - 9

CHAPTER II.

STEAM, SALT WATER, AND INCRUSTATIONS.

The Formation of Vapour and Steam—Boiling Point of Fresh and Salt Water—Analysis—The Causes which Influence the Boiling Temperature of Water—High Pressure Steam—Measure of Steam Pressures by Atmospheres—Steam when in Contact or not in Contact with Water—The Relation between Pressure, Density, and Temperature of Steam—Specific Gravity of Steam—Quantity of Water required to produce Condensation—Common and Superheated Steam—Analysis of Sea Water, - - - - - - 31

CHAPTER III.

Radiation, Oxidation, Etc.

The Radiation of Heat—The Absorption of Heat—Reciprocity of Radiation and Absorption—Good and Bad Radiators—Experimental Illustrations—Oxidation of Metals—Effects of Galvanic Action, - - - - 41

CHAPTER IV.

The Engine Before Watt, and Watt's Engine and Improvements.

Savary's Engine—Newcomen's Atmospheric Pumping Engine—Its Defects—The Discoveries of Watt—The Separate Condenser—The Expansive Working of Steam—Its Economy—Its Value in Regulating the Power of an Engine—Details connected with Watt's Single Acting Pumping Engine—The Steam Cylinder—Valves connected with Cylinder and their Action—The Condenser—The Air Pump—The Foot Valve—The Delivery Valve—The Snifting Valve—The Hot Well—The Piston-Rod—Connecting Rod and Crank—Stuffing Boxes and Glands—Parallel Motion—Method of Starting the Engine and of Regulating its Speed by the Governor—The Throttle Valve—The Cataract—Eccentric, - - - - - 45

CHAPTER V.

Beam Engine and Details.

Double Acting Condensing Beam Engine—Principle upon which it Works, etc.—Details of the Various Parts—Cylinder—How Constructed—Ports or Openings into the Cylinder, etc.—The Form of Slide Valve in Common Use—The Locomotive or Three-Ported Valve—The Lap on a Valve—The Eccentric—The Lead of a Valve—Cushioning the Steam—Clearance—Details of the Piston—Metallic Packing-Rings—The Expansion Valve and the Gear connected with it—The Supply of Water for Condensation—Blowing-through—Gauges for the Condenser—The Barometer Gauge—Method of Estimating Pressure by it—Errors in this

CONTENTS.

PAGE

Method, and Correction of the Same—The Fly Wheel—The Principle of an Equilibrium Valve—The Double Beat Valve—The Crown Valve—The Throttle Valve—The Gridiron Valve—The High Pressure Engine without Condensation—The Expansive Principle as Applied in the Double Cylinder Condensing Engine,   66

CHAPTER VI.

The Locomotive Engine.

History of Locomotive — Stephenson's Engine: The "Rocket" — General Description of a Locomotive—Crampton's Engines — Tank Locomotive — Bogie—Locomotive Boiler — Shell of Boiler — Through Tie Rods — Tubes — Clearance — Fire Box — Staying the Furnace—Fire Bars—Ash Pan—Smoke Box—Blast Pipe—Heating Surface—Safety Valves—Chimney—Damper — Steam Dome — Man Hole — Regulator—Whistle—Pressure Gauges—Salter's Spring Balance—Reverse Valve—Bourdon's Gauge—Sector, - - 90

CHAPTER VII.

The Water for a Locomotive.

Water Tanks — Water Cranes — Feed Pump — Giffard's Injector—Gauge Cock—Glass Water Gauge—Screw Plugs—Scum Cocks—Blow-Off Cocks—Heating Cocks,   114

CHAPTER VIII.

The Cylinders—Water Cocks—Grease Cocks—Piston and Piston Rod—Connecting Rod and Crank—Coupling Rod—Strap, Gib, and Cutter—Sector—Driving Wheel Tire—Counterweight to Wheels—Sand Cocks—Axle Boxes—Springs, Buffers, and Buffer Springs—Brakes—Draw Bar, - - - - - - - 125

CHAPTER IX.

The Road.

Tramway—Railroads—Curves—How the Carriages are Kept on a Curve—Rails—Fish Joints—Gradients—Ballast—Cuttings and Embankments—How Rails are

	PAGE
Laid — Two Ways — Broad and Narrow Gauge — To Adapt one Gauge to the other — Switches and Crossings — Tractive Force — Adhesion of Wheels to Rails,	137

CHAPTER X.

Indicator and Slide Diagram, - - - - - 148

PLATES.

I.—Section of Locomotive Engine, - - - -	94
II.—Fire Box, Fire Bars, Ash Pan, and Supports for Top of Fire Box, - - - - - - - -	100
III.—Plan of Cylinder and Driving Gear, - - -	125

STEAM.

CHAPTER I.

HEAT.

Definition—Expansion of Bodies by Heat—The Liquid and Gaseous States of Matter—Co-efficient of Expansion—Energy of Atomic Forces—Practical Illustrations—Temperature of Bodies, and Instruments for Measuring Temperature—Thermometers—Comparison of Thermometers—Graduation of Thermometers—Pyrometers—Capacity for Heat of Bodies—Calorimeter—Conversion of Heat into Work and Work into Heat—Consumption of Heat in Liquefaction and Vaporisation—Convection of Heat—Methods by which a Large Mass of Water may become Heated—Conduction of Heat—Good and Bad Conductors—Experimental Illustrations.

1. **Steam** is an elastic, invisible fluid generated from water by heat.

2. **Steam is Invisible.**—If we watch closely steam as it issues from a safety valve, a steam whistle, or even from the spout of a common kettle, we shall see nothing, or it is invisible. It is only at a distance from these orifices that it is rendered visible by parting with its heat to the air. When visible, properly speaking, it is no longer steam, but vapour; although it is not always wise to separate steam and vapour by such a hard and fast line. Some authorities say when water passes away insensibly without the mechanical application of heat it is vapour, but when heat is directly applied to the water it passes away as steam.

3. Steam is Elastic.—Take a cylinder or box, into which is tightly fitted a piston, and fill it with steam. If we now maintain the cylinder and steam at the same temperature, and apply a sufficient force to compress the steam into half the space, and then suddenly withdraw the force, the steam will again expand and fill the same space as before, driving the piston back again to its original position. The piston is returned to its place by the elastic force of the steam. Or we may illustrate the elasticity of steam much better thus: Suppose our cylinder full of steam, to be steam at a pressure of 15 lbs. on the square inch, and let the piston be at A B, and that from B to N be sixteen inches. If the piston be forced half-way down, or eight inches, to C D, then the steam, occupying one-half its former space, its pressure will be doubled, or on each square inch the pressure will be 30 lbs. Next force the piston to E F, four inches farther down, so as to reduce again the volume of the steam by one-half, or to compress it into one-quarter of its original volume, then the pressure will be again doubled, 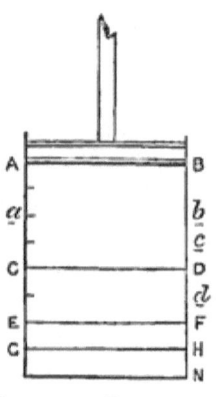 and will now be 30 × 2 or 15 × 4 = 60 lbs. on the square inch. If it be forced to G H, two inches still farther down, or the volume again decreased one-half, or occupying one-eighth of the original space, the pressure is now 60 × 2 or 15 × 8 = 120 lbs. on the square inch. We see by this illustration that the pressure increases as the space decreases. This is called Mariotte's or Boyle's law, and is generally expressed thus: *The temperature remaining the same, the volume of a given quantity of gas is in inverse ratio to the pressure which it sustains.*

4. Latent Heat.—The heat not sensible to the thermometer is termed latent heat or hidden heat.

5. Heat or Caloric.—When heat is applied to bodies

they immediately expand, and when cooled they contract. When the sun shines upon the air it expands, rises up, and causes currents of air or wind. When the sun shines upon the sea the waters expand at the equator, and flow towards the north and south.

When heat is applied to bodies the molecules immediately begin to oscillate or vibrate to and fro—the quicker the vibration the more intense the heat; as they cool they vibrate more slowly, or lose their motion; hence *heat is motion—the motion of atoms, and cooling is a loss of motion, or decrease of vibration.*

6. **The Liquid and Gaseous State of Matter.**—If sufficient heat be applied to the solid hard substance iron, it becomes a molten mass, and should a more intense heat be continued, the iron will pass off as an incandescent vapour or gas. The most familiar illustration we have of the liquid and gaseous state of matter is the common substance water. It is presented to us as the hard crystalline solid substance ice, as limpid water, and as gas in the form of steam, in each state it is endowed with perfectly distinct qualities. Our business is chiefly with those qualities presented to us when it is in the condition of a gas. Water, like every other substance (we will refer to the exceptions presently), expands by heat and contracts by cold. The liquid water itself expands by heat, but when the water is transformed to steam it occupies, in round numbers, 1700 times as much space, or, more exactly, 1669 times its volume. A cubic foot or cubic inch of water, evaporated into steam at the ordinary pressure of the atmosphere, fills a space equal to 1669 cubic feet or inches.

7. **Co-efficient of Expansion.**—Already it has been stated that all bodies upon being heated expand, and on cooling contract.

The amount a body expands in length, on receiving one degree of additional heat, is termed *the linear co-efficient of expansion.*

The amount the surface of a body expands, in receiving

one degree of additional heat, is termed *the superficial co-efficient of expansion.*

The amount a body expands in bulk, on receiving one degree of additional heat, is termed *the cubical co-efficient of expansion.*

The superficial co-efficient is generally double the linear, and the cubical three times the linear.

A gas or other elastic fluid, on being heated one degree centigrade, expands about $\frac{1}{273} = \cdot 003666$ of its volume, or $\frac{1}{490}$ for each degree Fahrenheit. Supposing we have 273 cubic inches of steam or gas in a vessel in which it can expand, upon the application of heat sufficient to heat it one degree centigrade, it will expand *one* inch, and occupy a space of 274 inches, heated two degrees it will occupy a space of 275 inches, three degrees, 276 inches, etc.

Co-efficients of Expansion.—The following are the co-efficients of a few well known substances:—

Substance.	Linear Co-efficient.	Cubical.
Zinc	·0000297	·0000890
Lead	·0000284	·0000890
Cornish Tin	·0000217	·0000690
Silver	·0000191	·0000574
Brass	·0000185	·0000554
Copper	·0000171	·0000512
Gold	·0000151	·0000453
Wrought Iron	·0000118	·0000354
Platinum	·0000088	·0000264
Glass	·0000087	·0000254

8. Bodies Expand by Heat and Contract by Cold.—The law is almost universally true, that bodies expand by heat and contract by cold.

(*a*) The most familiar illustration we have of this law is in the expansion and contraction of water when under the influence of heat and cold. Take water at a temperature of 4°C.; after the heat has been applied for a short time, it will begin to expand, and will continue to expand as the temperature increases, till it reaches the boiling point 100° C. After this, if we continue to apply heat, no alteration will take place in the temperature of the water. The additional heat that passes into the water is employed in converting the water into steam. A cubic inch of water will supply 1669 cubic inches of steam, or nearly a cubic foot. The result of another experiment was that a gallon of water, evaporated at 100°C., produced nearly 1700 gallons of vapour. When cold is applied to this vapour it contracts to its original volume.

(b) In building such bridges as the Albert Bridge, Saltash, and the Britannia and Conway tubular bridges, spaces are left for the expansion and contraction of the iron. The difference between the lengths of these bridges measured during the extreme heat of summer and the extreme cold of winter, is considerable.

(c) Experience has taught us that, in laying down the rails for a railway, spaces of about two-eighths or three-eighths of an inch must be left to allow the rails to expand in length. Were this not done, the molecular force of expansion would be sufficient to draw the spikes or lift the sleepers and rails out of their places.

Mr. Stephenson once stated that, in consequence of laying three or four miles of line, near Peterborough, with close joints, the heat of the sun on a warm day caused such an extension that the rails and sleepers were lifted in one place from the ballast so as to form an arch fifty feet long and three feet high in the air.

(d) The simplest plan to separate a crank from a shaft on which it has been shrunk; and, in fact, to disconnect any rust joint, is to apply heat, when the bodies (being of different dimensions) expand unequally and separate.

(e) Many other illustrations might be given, as, when warehouses constructed with fire-proof floors, etc., have been destroyed by fire, the walls of the buildings which were considered indestructible, have been thrown down by the enormous expansion of the iron girders, tie-beams, etc. Wheelwrights and carriage-builders, when they wish to place the tire upon a wheel, expand it by placing it in a fire, then slip it upon the wheel, and suddenly cool it, when the molecular power of contraction holds and binds the whole wheel firmly together.

9. Bodies Contract by Cold.—This may be illustrated by most of the foregoing instances of expansion by heat. A cubic foot of steam becomes a cubic inch of water when contracted by cold. The ends of railway rails are more widely separated in winter than in summer. This point will be further illustrated under the heading of Molecular Force; but a good illustration will be found in the method by which collars are shrunk on a shaft. A neat way of putting collars on heavy marine shafts where the journals come, is this—bosses are turned on the shaft, and two ribs, three or four sixteenths of an inch high, are left on the bosses for the collars, which must be prepared in the lathe, and heated and slipped over the ribs, then, upon contracting with the cold, they will firmly grip the shaft.

10. The Enormous Power of Expansion and Contraction.—When bodies expand, the molecules of which they are composed are pushed farther asunder by the oscillatory motion communicated to them. The heat may be described as entering the substance, and immediately setting to work to separate the particles. The power or energy they exert to do this is immense. The following are illustrations of the energy of molecular forces. We have already mentioned several under the heads Expansion and Contraction :—

(*a*) When a dry wooden wedge is driven into the crevice of a rock, and moistened with water, the wedge swells and splits the mass. Thus many accidents have happened to grinders through the wedges swelling between the axle and the stone, and causing the latter to burst. Of course, in this case, centrifugal force assisted the wedges.

(*b*) When a rope is moistened, the diameter becomes larger, and the rope shorter, for the fibres are drawn in by this enlargement. It is said that, in lifting the statue of Nelson into its place in Trafalgar Square, the ropes had stretched through the great weight, and the blocks were close to each other. The whole operations would have failed, although the hero was within a very short distance of his place, had not a sailor cried out, "Wet the ropes." The hint was immediately taken, and the work accomplished.

(*c*) Water is turned into steam by heat; this heat endows the water with (atomic) force sufficient to drive the locomotive, to propel the steamship round the world, to work the mill, the forge, the hammer, the pump, etc.

(*d*) If the wall of a large building be bulging out, and an iron girder placed in a proper position, the power of contraction by cold will subserve the purpose of bringing it into the perpendicular. It has been done on a large scale in France. A girder (or girders) was fitted across the building, with strong wall plates at each end, and screwed up as tightly as possible. All along the girder was applied a number of gas jets, and as it expanded by the heat, the screws were tightened. The girder was then allowed to cool, and the strain of its contraction was sufficient, after repeating the process several times, to draw the walls into the perpendicular.

(*e*) We may also add, that the Gulf Stream and the trade winds are caused by the atomic force of heat (but see Convection).

11. Molecular Force, or Atomic Force.—All molecules are under the influence of two opposite forces. The one, *molecular attraction*, tends to bring them together; the other, *heat*, tends to separate them; its intensity varies with its velocity of vibration. *Molecular attraction* is only exerted at infinitely small distances, and is known under the name of *cohesion, affinity,* and *adhesion.*

By the force of *cohesion* everything is held together. Heat and cohesion are antagonistic to each other. When heat predominates in liquids they become gases, when cohesion solids.

Chemical affinity is a form of molecular force that greatly assists the engineer; by chemical affinity the various products of combustion in the air and the fuel combine, and in the act of combination produce the heat necessary for his purpose.

Adhesion is the molecular force exerted between bodies in direct contact. When two pieces of lead have their pure metallic surfaces laid bare, and are put together with pressure and a twist, they become united so as to require considerable force to separate them. When we come to speak of the marine engine, and the way in which the thrust of the shaft of the screw is received upon the thrust block, reference will be made to the method first adopted to receive the thrust, namely, on a series of discs, which sometimes became so clean for want of attention to the lubrication, that the pure metallic surfaces firmly united by the molecular force of adhesion, and the shaft broke at a distance from the discs.

Heat is another atomic force, which, by causing bodies to expand and contract, exerts enormous power as already illustrated.

12. Temperature of Bodies.—Temperature of a body is the measure of the intensity of heat in the body. A body may contain a large amount of heat that is not sensible to a thermometer.

13. Thermometer is an instrument for measuring the temperature or intensity of heat in a body. It is constructed on the principle that bodies expand and contract when subject to cold and heat with a certain amount of regularity within certain limits. The ordinary thermometer consists of a closed glass tube with a capillary bore, terminating at the lower end in a bulb, which, with the bottom part of the tube, contains mercury or spirits of wine, etc., the rest of the tube being a vacuum. A graduated scale by the side of the tube indicates the amount of expansion or contraction of the mercury. There are three kinds of thermometers.

(1) Fahrenheit's, chiefly used in England and America.

(2) That of Celsius, chiefly used by the French.
(3) That of Reaumur, used by the Germans.

14. (1) **Fahrenheit's Thermometer.**—The increment of expansion by heat and contraction by cold of mercury is practically the same for all temperatures for which a thermometer can be employed. Hence mercury is much better adapted for a thermometer than water or spirits of wine. Fahrenheit named the greatest degree of cold attainable in his time by artificial means 0° F., and the freezing point 32°. Hence the graduation of his thermometer commences at 32° below the freezing point, and between freezing and boiling there are 180°, so that the boiling point is 212° F.

15. (2) **Centigrade Thermometer.**—The freezing and boiling points in the centigrade thermometer are 0° and 100° respectively. The method of indicating the measure of heat, termed centigrade, is found so convenient that it is fast superseding Fahrenheit.

16. (3) **Reaumur**, or Romer, introduced a much more arbitrary division of the scale. He called the freezing point 0°, the boiling point 80°.

We now see that in Fahrenheit's scale there are 180° between the freezing and boiling points, in the centigrade 100°, in Reaumur 80°.

Rules to compare the reading of one thermometer with that of another:—

(1) To convert Fahrenheit's degrees to centigrade—
    Subtract 32°, then multiply by 5, and divide by 9.
(2) To convert centigrade to Fahrenheit—
    Multiply by 9, divide by 5, and add 32°.

(3) To convert centigrade to Reaumur—
  Multiply by 4 and divide by 5, or subtract one-fifth.

(4) To convert Reaumur to centigrade—
  Multiply by 5 and divide by 4, or add one-quarter.

(5) To convert Fahrenheit to Reaumur, or Reaumur to Fahrenheit—
  First bring them into centigrade, then reduce to Fahrenheit or Reaumur, whichever may be required.

Exercises on the reduction of the number of degrees of one thermometer to an equivalent number of another, will be found at the end of the chapter.

17. Pyrometers.—Pyrometers are used for measuring intense temperatures. It is evident to the most casual observer that the thermometer will measure a degree of heat but little beyond the temperature of boiling water. To measure the intense heat of the kiln of the porcelain manufacturer, the puddling furnace, the blast furnace, the boiler furnace, flues, etc., requires instruments of perfectly different construction. These are found in such as Daniell's Pyrometer, Wedgewood's, the Sevres, Lavoisier and La Place's, Houldsworth's, etc.

Daniell's Pyrometer consists essentially of a small bar of platinum and a scale. A solid bar, in length about eight inches, is cut out of a piece of black lead earthenware, down its centre is drilled a hole reaching nearly to the bottom. Into this is inserted a tube of platinum reaching down to the end of the hole, leaving room at the top to allow a small tube of porcelain to be placed in, and to touch the end of the platinum. This tube of porcelain is called the *index*. The whole is named the *register*. When it is desirable to ascertain the temperature of a heated body, furnace, etc., this is placed within the heat, and sufficient time allowed for it to acquire the same temperature. When it is withdrawn, it is found that the heat has expanded the platinum, which, in its turn, has driven out the porcelain tube a certain space, according to the intensity of the heat; the porcelain is prevented from returning by a platinum strap.

The *Register* is, after it has cooled, next applied to a scale properly graduated, to enable the observer to read off easily the change in temperature.

18. Unit of Heat.—A unit of heat is the amount of heat necessary to raise the temperature of a pound of water one degree. Suppose a pound of water to be raised from 10°C. to 20°C., it has received ten additional units of heat; if five pounds of water be raised 5°, each pound has received five units of heat, and the whole twenty-five units. If we raise the temperature of half-a-pound of water 10°, we communicate to it *five* units of heat.

19. Capacity for Heat of Bodies.—The capacity for heat of bodies means their power of storing up heat. To work the same change of temperature in different bodies requires different amounts of heat. A given quantity of heat put into one body will cause a greater amount of motion than when put into another. Suppose, for instance, we throw six balls, the same size, of silver, tin, bismuth, copper, lead, and iron into boiling water, each will soon acquire the temperature of the boiling water, 100°C.; now take them out of the water, it will be found that you can almost at once handle the bismuth and lead, soon after the tin, then the silver, last of all the copper and iron remain hot the longest. The reason is this, that to raise the lead and bismuth to the temperature of 100°C. requires much less heat than to raise the tin. Tin requires less than silver, silver less than copper, and iron more than either; and, therefore, having more heat it takes a longer time to lose the motion. If we had taken the same balls and put them upon a thin cake of wax after heating them in boiling oil or water, it would be found that the iron would melt and fall through the wax first, the copper next, silver next, perhaps, if the wax were sufficiently thin, while the lead and bismuth would not get through at all. The reason is this, the iron has the greatest specific heat, or it has stored up more heat than the others, and, therefore, it has enough to impart to the wax to melt it. The same with the copper, while the bismuth and lead, having a less capacity for heat than the others, they have less to give up, or less motion to impart to the wax to melt it and work their way through.

20. The Calorimeter is not used to measure the temperature of a body, but to ascertain the total amount of heat in it, or to find the specific heat.

Two similar metallic vessels are placed, one within the other, so as to leave a space between them. This space is filled with pounded ice, while a discharge pipe proceeds from the bottom of the external vessel to carry off all water that may be produced through the liquefaction of the ice by the external air. A third, and nearly similar vessel, is placed within the second, leaving a space between it and the second vessel, which is also filled with pounded ice; a second discharge pipe (with a stop cock) proceeds from the second vessel without communicating with the outside one. Each vessel is provided with its proper cover. It is obvious that the ice in the inner space cannot be affected by the temperature of the external air when the calorimeter is closed. The substance, whose specific heat we wish to ascertain, is placed, after observing its temperature, within the third or inner vessel. It is perfectly clear that any heat the body may contain, will communicate or lose its motion to the ice in the *second* space, or the ice will take up the heat from the substance as latent heat, and become converted into water; this is then allowed to pass through the discharge pipe leading from the inner vessel, and is collected. This water will at all times be proportional to the heat stored up in the given substance placed within the calorimeter.

By the calorimeter, it has been ascertained that to raise the temperature of water 1°, requires thirty times as much heat as would be required to raise mercury 1°. Or the same heat that would raise 1 lb. of water 1°, would raise the temperature of 30 lbs. of mercury 1°; and this is what is meant when we say the specific heat of mercury is $\frac{1}{30}$ or ·03 that of water. Iron requires $3\frac{2}{3}$ more heat than lead to work in it the same change of temperature; practically, this means that lead will heat $3\frac{2}{3}$ times quicker than iron; at the same time it will cool very much more quickly than iron. It is obvious that to heat 2 lbs. of water 1°, requires twice as much heat as to heat 1 lb. of water 1°. The relative quantity of heat necessary to produce the same change of temperature in different bodies is their specific heat. We said the capacity for heat of water was thirty times that of mercury; hence this latter substance is so well adapted for thermometers; we see at once how sensible it must be to the least accession or subtraction of heat. Again, the capacity for heat of air at constant pressure, is about one quarter that of water, or more accurately ·237; hence 1 lb. of water, whose specific heat is 1, on losing 1° of heat, will increase the temperature of ($\frac{1}{·237} =$) 4·2 lbs. of air 1°. But water is 770 times heavier than air. Hence if we compare volume instead of weight, a cubic foot of water, on losing 1° of temperature, will increase that of $770 \times 4·2 = 3234$ cubic feet of air 1°.

Capacity for heat may be defined as the quantity of heat necessary to raise the *same weight* of different substances through the same *number of degrees* of temperature; but it must not be defined as the amount of heat necessary to raise a pound weight of a given substance one degree in temperature, or else we shall confound it in the case of water with the unit of heat. Capacity for heat is found thus: one, two, three pounds, ounces, etc.; any weight may be chosen, of any substance, and heated so many degrees, one, two, three, etc. (generally heated in boiling water), and then put into the calorimeter, then according to the quantity of ice melted we have the capacity for heat. The quantity each substance liquefies is noted, the whole compared with water as a standard, and the capacity for heat determined.

The following are the specific heats, or capacity for heat, of a few well known substances:—

Bismuth,......·0308	Copper,.........·0949	Air, ...·237
Lead,...........·0314	Iron, ...·1098	Steam,......·4805
Mercury,......·0333	Glass, ...·1770	Ice,..........·504
Platinum,.....·0355	Sulphur, ...·1844	Water, ...1·000
Silver, ...·0557	White marble,·2158	

21. Convection—Methods by which Large Masses of Air or Water become Heated.—"Convection is the transfer of heat by sensible masses of matter from one place to another." Water can only be heated by convection; it is scarcely possible to heat it by conduction. Our rooms are ventilated by convection, smoke ascends the chimney by the same principle, and all our winds and currents, in both air and water, are caused by this convection. The wind-sails of a ship afford an instance in which this law of nature is made available for ventilation.

If A B be a glass vessel or large Florence flask filled with water, when heat is applied at A, the water near A is immediately heated and expanded, and becoming specifically lighter rises up, and the colder water from above falls down to supply its place; this continual change goes on as long as the heat is applied at A, and is called convection. If a little cochineal be placed in the water, it will sink to the bottom of the flask, and heat being applied as before, the cochineal directly leaves the bottom, ascends up the middle, and then descends by the sides, returning again to the heat. By this simple experiment the action of convected water is made visible to the eye.

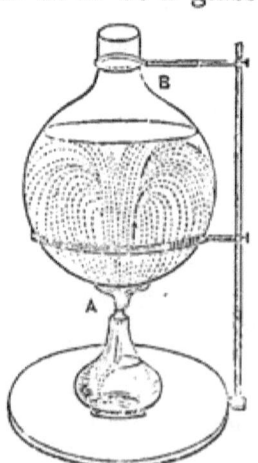

CONVECTION OF WATER.

Let C D be a large test-tube filled with water, and held by an holder in the position indicated by the figure; then let heat be applied at D, it will be found almost impossible to heat the water in the test-tube, for the heated or convected water rises perpendicularly up from the heat, confining itself to the top of the tube, and scarcely any heat is *conducted* downwards, for, of course, the convected or light water cannot run down, or mix itself with, or rather communicate its motion to, the heavier water below. Large masses of water can only be heated by convection, and therefore all furnaces should be placed as low down in the boilers as possible, while below the bars there should be but little if any water.

CONDUCTION OF WATER.

A patent fire-door is used for boilers, which is nothing but the application of the principle of convection : the doors are made with front and back plates, and hollow within. In the front plate are a few openings, one to one and a half inches in diameter, the back plate is thoroughly perforated with smaller holes. The air goes in at the bottom of the front plate, and out at the top, carrying off the heat, so that the front of the door is never heated to redness, the current of convected air carrying off the heat. In precisely the same way the funnels of steamers are kept cool, and passed through the wood of the decks. A casing is placed entirely round the funnel, passing into the engine-room, and sometimes spreading out over the boilers. A stream of air then continually runs up between the funnel and the casing ; this air takes the heat out of the funnel as it passes upwards, and keeps it from becoming too hot. Holes are often made at the bottom of the casing for the passage of additional air.

22. **Latent Heat of Water (or Ice).**—If a pound of ice at 0°C., be mixed with a pound of water at 79°·4C.,

the water will gradually dissolve the ice, being just sufficient for that purpose, and the residuum will be two pounds of water at 0°C. The 79·4 units of heat which are apparently lost, have been employed in performing a certain amount of work, *i.e.*, in melting the ice or separating the molecules and giving them another shape, and as all work requires a supply of heat to do it, this 79·4 units have been consumed in performing the work necessary to melt the ice, and is termed the **Latent Heat of Water.** If the pound of water were re-converted into ice, it would have to give up the 79·4 units of latent heat; hence we see why it should be called the latent heat of water, and not the latent heat of ice. The three forms of water are, then, (1) a solid, as ice; (2) a liquid, as limpid water; (3) a gas, as steam.

23. Latent Heat of Steam—The latent heat of steam at a pressure of 15 lbs., or thirty-two inches of mercury, is 537°·2C. We will describe an experiment which will help to illustrate this point, and fix the fact in the memory. Suppose that we have two very small vessels connected at their tops by a tube. Let one contain a pound of water, at the temperature of 0°C., and the other five and a half pounds, at the same temperature. If a spirit lamp be applied beneath the vessel containing the one pound of water, its temperature will gradually rise to 100°C., when ebullition will begin, and if the heat be continued, the water will not increase in temperature, but will pass off as steam along the tube to the second vessel, where the five and a half pounds of cold water will condense the steam and absorb the heat, which first enters and passes from the one pound, as long as the spirit lamp is applied to it. This operation of condensation and absorption will continue until the one pound of water is all converted into steam and re-converted into water. At the moment that the evaporation of the pound of water is completed, the heat transferred by the steam from one vessel to the other will cause the five and a half pounds of water to boil. It will be found that there are now in the second vessel six and a half pounds of water, at a temperature of 100°C. As the 1 lb. takes 100 units of heat to make it boil, the $5\frac{1}{2}$ lbs. take $5\frac{1}{2} \times 100 = 550$ units; or, as there are $6\frac{1}{2}$ lbs. of water in B, the total quantity of heat is $100 \times 6\frac{1}{2} = 650$ units of heat. The boiling water, the one pound, never exceeded a temperature of 100°, all the rest of the heat went to evaporate the water; and as we know there are 650 units of heat in the $6\frac{1}{2}$ lbs., therefore the latent heat of steam is 550. Exact experiments make the $5\frac{1}{2}$ lbs. 5·372. Hence the latent heat deduced from the experiment will be $5·372 \times 100 = 537·2$. This 537°·6 C., or 966°·6 F., is the latent heat of steam. In making the experiment, ounces or smaller quantities of water are employed, and not pounds.

24. Consumption of Heat in Liquefaction and Vaporisation.—This is but another way of putting the facts connected with the latent heat of water and steam. We have seen that the latent heat of water is 79°·4 C., or to liquefy a given quantity of ice requires this amount of heat; to raise the water to its highest temperature consumes 100°C. more; next, to vaporise it consumes 537°·2 C.

When heat is imparted to a body its atoms push each other asunder, and the molecules commence to oscillate more or less rapidly. The more intense the heat the quicker the particles oscillate; by raising the temperature you increase the oscillations, while cooling is a decrease of vibration, or loss of motion.

25. Conduction.—If we place one end of a poker or piece of iron in the fire, the molecules of the iron in the fire immediately begin to oscillate, and each molecule strikes its neighbour, passing the motion on, so that the end of the poker out of the fire also becomes warm. The process by which the heat is passed up the poker is called conduction. There are good and bad conductors. The metals are generally good conductors, and the earths, sawdust, ashes, stone, glass, chalk, etc., bad conductors. Silver is one of the best conductors. If we call its power of conduction 100, that of copper is 74, of gold 53, iron 12, lead 9, bismuth 2. A knowledge of this property of heat will teach an engineer on what to bed or surround his boiler, so that the least possible heat may be conducted out of it; also, in what he may case his steam pipes, cylinder, etc., to attain the same end.

The following are a few simple illustrations of the principle of conduction:—If a short piece of iron and a piece of glass the same length and size be placed in the fire, a little time afterwards we may handle the glass at the farther end, but not the iron, because the motion of the heat has passed up the iron more readily than up the glass. Again, if we obtain two short bars, one of platinum and another of tin, and fasten along them, by means of wax (which melts at $16°\frac{1}{5}$C.), a row of small balls, then bring the two ends together in the flame of a burner, so that they shall be subject to the same heat, we shall find that the balls will drop off three times faster from the platinum rod, by means of its superior conductibility, than from the tin. The same experiment can be tried with rods of copper and iron, when they will fall more rapidly from the better conductor of the two, copper. As a rule, conduction is most strongly exhibited by solids, particularly by metals. Dense bodies (not including earths) are the best conductors, and light and porous ones the worst—the latter being non-homogeneous and not capable of communicating the motion or passing it on. Feathers, down, flannels, fur, blankets, woollens, etc., keep us warm because they are bad conductors, and will not allow the motion to pass from the body. On the same principle a few sheets of newspaper placed on a bed keep the sleeper warm. Liquids and gases are generally very bad conductors of heat.

26. Conversion of Heat into Work and Work into Heat.—Loss of heat is loss of work. It is the province of the engineer to convert heat into work, and to let none escape until it has done its duty. The most familiar illustration we can give is the locomotive. Fire is put under the boiler, the heat converts the water into steam, which drives the engine and train; before the train can be stopped the whole moving force must be destroyed, which is effected by shutting off the steam and putting on the brake, when the force that moves the train is re-converted into heat, and sparks and fire issue from the brake. So that the work, or moving force, is turned into heat. When a carriage or waggon is going down a hill, the drag is put on to destroy the moving force, which it does by converting it into heat, and making the drag very hot. We must remember, at the same time, that the horses drawing the carriages convert the heat of their bodies, supplied and constantly renewed by the food they eat, into work, part of which work we have just shown is re-converted into heat in the drag in going down hill.

QUESTIONS AND MATHEMATICAL ILLUSTRATIONS FROM EXAMINATION PAPERS.

1. Convert $18°\cdot5$ C. to Fahrenheit (1866).
 Since $100°$C. $= 180°$F. (between freezing and boiling)
 $\therefore 5°$C. $= 9°$F.
 $\therefore 18°\cdot5$C. $= 65°\cdot3$F.

$$\begin{array}{r} 9 \\ \hline 5)166\cdot5 \\ \hline 33\cdot3 \\ 32 \\ \hline 65°\cdot3\text{F.} \end{array}$$

add so that the starting point may be 32° below freezing.

2. What degree on the centigrade thermometer corresponds to 158° on Fahrenheit's?

By the same reasoning as in the previous question, but subtracting 32° to start from the freezing point,

$$158°\text{F.} = 70°\text{C.}$$

$$\begin{array}{r} 32 \\ \hline 126 \\ 5 \\ \hline 9)630 \\ \hline 70°\text{C.} \end{array}$$

3. Reaumur's scale shows 45°, what is the corresponding temperature (1) on the centigrade? (2) on Fahrenheit's scale?

Centigrade.	Fahrenheit's.
80° R. = 100° C.	Instead of converting the 45°
∴ 4° R. = 5° C.	R. to F., we will convert to
∴ 45° R. = 56¼ C.	56¼ C. to F.
	56¼ C. = 133¼ F.

$$\begin{array}{r} 5 \\ 4\overline{)225} \\ \hline 56\tfrac{1}{4}\ C. \end{array}$$

$$\begin{array}{r} 9 \\ 5\overline{)506\tfrac{1}{4}} \\ \hline 101\tfrac{1}{4} \\ 32 \\ \hline 133\tfrac{1}{4}\ F. \end{array}$$

4. The linear co-efficient of expansion of iron is ·0000123 for 1° C. I have a rod of iron 20 feet long, and heat it from 20° C. to 290° C., what is the increase in length?

The temperature is raised 290° − 20° = 270°

For 1° C. each foot of iron increases ·0000123 feet in length
For 270° C.   ,,   ,,   ,,   ·0000123 × 270   ,,
For 270° C. 20 feet ,,   ,,   ·0000123 × 270 × 20 ,,
      = ·06642 feet = ·79704 inches, or over half an inch.

5. The co-efficient of expansion of gas is $\tfrac{1}{273}$ for one degree centigrade, find the increase in volume of 100 cubic feet of gas heated from 10° C. to 100° C.

The temperature of the gas is raised 100° − 10° = 90° C.

1 cubic foot on being heated 1° C. increases $\tfrac{1}{273}$
1 cubic foot   ,,   ,,   90° C.   ,, $\tfrac{90}{273}$
100 cubic feet ,,   ,,   ,,   ,, $\tfrac{90 \times 100}{273}$
            = 33 nearly

∴ Volume of the 100 cubic feet after being heated = 100 + 33 = 133 cubic feet.

6. Convert 100° C. to Fahrenheit.   *Ans.* 212° F.
7. Express 28° F. in degrees centigrade.   *Ans.* −2⅔ C.
8. What degree centigrade and Fahrenheit corresponds to 80° R.?   *Ans.* 100° C., and 212° F.
9. What degree centigrade corresponds to 100° R.?
     Ans. 125° C.
10. The linear co-efficient of expansion of copper is ·0000171. How much will a rod 10 ft. long increase in length if heated from 0° C. to the temperature of boiling water?
     Ans. nearly ¼ inch.
11. A brass letter is attached to the glass of a window, find how much the amount of expansion varies between the extreme heat of winter and summer, extreme of winter being −3° C., extreme of summer 35° C.; the co-efficient of expansion of glass is ·00000876, and of brass ·0000185.   *Ans.* ·0044144 inches.

12. The temperature of 20 cubic feet of gas is increased from 14° C. to 49° C., find the increase in volume and present volume.
Ans. 2·56 and 22·56 cubic feet.

13. If 40 volumes of gas have their temperature raised 60° C., what is the increase?   *Ans.* 8·79 volumes.

14. Explain what is meant by capacity for heat and latent heat. What is the latent heat of steam at the ordinary atmospheric pressure (1863)?

15. Distinguish between conduction and convection. Mention some substances that are bad conductors, and state to what uses they are applied in the steam engines (1863).

16. Describe the several methods by which heat is propagated. Explain the terms capacity for heat and latent heat. What is the latent heat of steam (1864)?

17. What is meant by temperature? What are the general effects of adding heat to or subtracting it from a body (1865)?

18. A centigrade thermometer marks 5°, what will a Fahrenheit thermometer mark (1865)?   *Ans.* 41° F.

19. Describe Daniell's pyrometer. For what purpose is it used (1865)?

20. Define capacity for heat, latent heat, and unit of caloric (1865).

21. Show how to graduate a thermometer.

22. Why is it necessary to take the height of the barometer in account in determining the boiling temperature (1866)?

23. Show how to convert degrees on a centigrade into degrees on Fahrenheit's scale.

24. What temperature F. corresponds to 49°·5 C. (1866)?
Ans. 121°·1 F.

25. What is meant by latent heat? Show under what circumstances heat becomes latent (1866).

26. What do you understand by conduction and convection as applied to heat (1867)?

27. What is the latent heat of steam? How is its amount ascertained (1867)?

28. What is the distinction between sensible and latent heat? Describe an instrument for measuring the former (1868).

29. Under what circumstances generally (1) does heat become latent? (2) does latent heat become sensible?

30. What amount of latent heat becomes sensible when ice is thawed into water (1868)?

31. Show how a thermometer is graduated. Compare the graduations on Fahrenheit's, Reaumur's, and the centigrade scale. Reaumur's scale shows a temperature of 15°, what will the centigrade and Fahrenheit's scales respectively show for the same temperature (1868)?   *Ans.* 18°¾ C. and 65°¾ F.

32. What do you understand by the conduction of heat?

Mention one or two good, moderate, and bad conductors of heat (1869).

33. How can it be shown that the temperature at which water boils depends upon external pressure? What is high pressure steam (1869)?

34. Under what circumstances does heat become latent?

CHAPTER II.

STEAM, SALT WATER, AND INCRUSTATIONS.

The Formation of Vapour and Steam—Boiling Point of Fresh and Salt Water—Analysis—The Causes which Influence the Boiling Temperature of Water — High Pressure Steam— Measure of Steam Pressures by Atmospheres—Steam when in Contact or not in Contact with Water — The Relation between Pressure, Density, and Temperature of Steam— Specific Gravity of Steam—Quantity of Water required to produce Condensation—Common and Superheated Steam— Analysis of Sea Water.

27. Vapour and Steam.—Steam was defined as an elastic invisible fluid, produced from water by the application of heat. So long as it is invisible some authorities count it steam, and as soon as it becomes visible they call it vapour. Others, again, give the term steam to all vapour produced artificially by heat; when water passes away insensibly, as through the influence of the sun or air, it is called vapour. *Evaporation* is the act of converting water into vapour. *Liquefaction* is the act of converting a solid into a liquid, or a gas into a liquid. Ice is converted by heat from a solid state to a liquid condition. Heat is the sole agent in the liquefaction of solids; on the contrary, cold will liquefy certain gases, while it will render some liquids, as water and mercury, solids.

28. Boiling Point of Fresh and Salt Water.—Speaking of the thermometer, it was stated that 100°C., 80°R., and 212°F., marked the boiling point of water respectively on each system of graduation. The boiling point of a liquid is defined as "that temperature of the liquid at which the tension of its vapour overcomes the resistance of the atmosphere." At the sea level, fresh water is found to boil at a temperature of 100°C., and for every 1062 feet that we ascend in vertical space, water will boil at a temperature of 1°C. less, the barometer standing in such cases at about 30 inches of mercury. The reason why water will boil at a lower temperature as we ascend, is, that the tension of its vapour meets with less resistance from the pressure of the atmosphere, and therefore *ebullition* takes place earlier. Sea water is heavier than fresh, in the proportion of 1 : 1·024. In consequence of this, salt water boils at a higher temperature than fresh water, because the tension of the vapour has a greater resistance to overcome in separating itself from the water. Sea water with $\frac{1}{30}$ of salt in it, at a pressure of 15 lbs. on the square inch, boils at a temperature of 100°$\frac{2}{3}$C., with $\frac{2}{30}$ of salt it boils at 101°$\frac{1}{3}$C., with $\frac{3}{30}$ at 102°C., etc. Thus the saltness of the water influences the temperature at which sea water boils. As the boiling point varies with the pressure of the atmosphere, so precisely in the same way the pressure of the steam upon the surface of the water in a boiler will have a tendency to raise the boiling point, because the tension of the vapour has a greater pressure or resistance to overcome before it can free itself from the water. When fresh water is used in a boiler, if we know its temperature we can tell the pressure, and if we know the pressure we can tell the temperature.

TABLE OF PRESSURES AND TEMPERATURES.

Pressure in lbs. per sq. in.	Temperature Centigrade.	Pressure in lbs. per sq. in.	Temp. C.	Pressure in lbs. per sq. in.	Temp. C.
15 lbs.	100°	55	142°·4	100	165°·3
20 lbs.	109°	60	145°·6	105	167°·3
25 lbs.	116°	65	148°·4	120	172°·9
30 lbs.	121°·7	70	151°·2	135	178°·
35 lbs.	126°·8	75	153°·8	150	182°·6
40 lbs.	131°·$\frac{1}{3}$	80	156°·3	165	187°·
45 lbs.	135°·4	85	158°·8	180	190°·9
50 lbs.	139°	90	161°	195	194°·6

The lesson to be learnt from the foregoing table is this, that if we find that the pressure gauge of a boiler using fresh water stands at 40 lbs., then the temperature of the water in the boiler is $131°\tfrac{1}{3}$ C.; or, on the contrary, if by a thermometer we can ascertain that the temperature of the boiler water is $131°\tfrac{1}{3}$ C., we know that the steam pressure *in the boiler* must be 40 lbs.

29. Salt in Sea Water and in a Marine Boiler.— As water evaporates, the steam passes away, leaving all impurities behind—whatever solid substances may be in solution in the water they remain in the boiler; hence a marine boiler using sea water becomes a large evaporating salt-pan, unless steps be taken to get rid of the salt.

30. Analysis of Sea Water.— The principal substance which is held in solution by sea water is chloride of sodium, or common salt. The following may be taken as a fair analysis of sea water: Out of every hundred parts of solid matter in sea water

```
 72 parts are sodic chloride
 11   ,,    ,,  magnesic chloride
  6   ,,    ,,  magnesic sulphate
  5   ,,    ,,  calcic sulphate
  2   ,,    ,,  calcic carbonate
  4   ,,    ,,  organic and other unimportant substances.
---
100
```

The substances most injurious to a marine boiler in the above, are the calcic sulphate and calcic carbonate, the second forms solid incrustations on the boiler, while the other eats it away. The amount of salt, it is seen, is 72 parts out of one hundred. Again, if thirty gallons of sea water be taken and evaporated, the residuum will be a gallon of salt; so that when we consider the large amount of water used by an engine, we can soon form an idea of the quantity of salt, etc., left behind, and how quickly danger may ensue through carelessness or want of proper precautions.

31. Point of Saturation.— It has been a problem for

the solution of marine engineers how to best get rid of these impurities, especially the salt sodic chloride, as it is present in the largest quantities. The only effectual method of disposing of the salt is not to let the water in the boiler become too salt. This is done by opening a communication, called the blow-out valve, between the boiler and the sea, and letting the water run from the boiler into the sea, and then filling up the boiler with fresh sea water. Sea water contains $\frac{1}{30}$ of salt, etc. Suppose we have a boiler containing a thousand gallons of salt water, and one half, or 500 gallons, is evaporated, it is evident that the remainder will contain double the original quantity of salt, or $\frac{2}{30}$. If one half of this be evaporated, so as to leave only 250 gallons, these will contain $\frac{4}{30}$ of salt, after which, if no feed enter the boiler, the water will soon reach what is called the point of saturation, $\frac{11}{30}$, and the salt in the water will fall down, for the water is *saturated* with salt, or will hold no more. The usual practice is never to allow the salt in the boiler to exceed $\frac{4}{30}$, or to blow out from the boiler *one-third* the quantity evaporated. If 300 cubic feet are evaporated, another hundred must be blown out, or as the 300 + 100 must be equal to the total amount of feed water, of the latter one quarter must be blown out. It may be illustrated thus: Of every 100 gallons or cubic feet of water that enter a marine boiler, 75 may be turned into steam, and the remainder blown out of the boiler into the sea. Such a practice will prevent the boiler water exceeding $\frac{4}{30}$ of saltness.

32. Brining the Boiler.—The process of blowing the brine out of the boiler is frequently called "brining the boiler." It is a custom to blow out at certain given periods, or after the engine has been at work, every 3 or 4 hours; but it is evident that if we have intelligent men driving our engines, they ought, by examining the water, or by calculating how much feed they have used, to know better than to depend upon such a rule of thumb method when to brine the boiler.

33. Incrustation of Land Boiler.—Land boilers become incrusted by the deposition of lime, iron, and other injurious substances upon the inside of the boiler. Just as we see a kettle incrusted by "fur," so land boilers are liable to the same thing, but to a much greater extent. Water from off the oolite, chalk, and other formations containing lime always incrusts boilers.

34. To Prevent Scale Forming in Boilers.—It has already been shown how this is accomplished in marine boilers. In land boilers nothing will prevent the formation of scale but the use of absolutely clean water. Many empirical receipts are given, but the most they do is to throw down the substances held in solution in the form of powder, which is simply a change of evils without any compensating advantage.

35. To Clean Scale from a Boiler.—This should always be done with the chipping chisel. It is condemned by many engineers as a vicious and bad custom, to throw a few shavings into the boiler, which, by creating sudden heat when set on fire, causes the inside shell of the boiler rapidly to expand, and thus separate itself from the incrustation.

36. Danger of Scale.—This will be fully treated under the head of boiler explosions.

37. High Pressure Steam Does not Scald.—If steam at high pressure be issuing from an orifice and the hand be placed in it, it will not be scalded. The reason must be that, as it issues into the air, the pressure is decreased and reduced to 15 lbs. The steam, therefore, immediately takes to itself the deficient latent heat from the air; for the latent heat of high pressure steam is less than that of low pressure, and as steam, when it has issued into the atmosphere, is under a pressure of only 15 lbs. on the square inch, it must necessarily take to itself the required amount of latent heat for that pressure. If the pressure had been 30 lbs., the deficient latent heat would have been 22°C. The steam is, therefore, busily employed in taking these 22° of heat from the atmosphere, and even from the hand placed in it; and so, under the circumstances, will rather cool than scald the hand.

38. Measure of the Pressure of Steam.—The pressure of steam is measured by atmospheres. Steam of 15 lbs. pressure is steam of one atmosphere, of 30 lbs. pressure of two atmospheres, etc. It is frequently used as high as six or seven atmospheres; but even ten, or 150 lbs. pressure, is employed. Steam below two atmospheres is termed *low pressure steam*, and all pressures above, *high pressure steam*. This is one of the points, and really of no consequence, upon which scientific men have no uniform opinion; what some consider high pressure, others consider low or ordinary pressure steam, while some term steam above one atmosphere high pressure. It is better, in the unsettled state of opinion, to make no distinction, as above, between high and low pressure steam, but simply state that you are using steam of one, two, three, or more atmospheres.

39. Superheated or Surcharged Steam.—It has become a practice to allow the steam, before it enters the cylinder, to pass from the boiler into a series of tubes, or into a strong iron chamber in which a large quantity of vertical or horizontal tubes are fitted; in these the steam is further heated to increase its elasticity by the heat that is passing away up the funnel or stack, thus from a given quantity of steam a maximum amount of work is obtained with a minimum amount of fuel consumed. A simple method of superheating steam is to allow it to pass from the boiler by means of a short pipe into a small chamber round the bottom of the funnel, and thence let it pass to the engine.

40. The Advantage of Superheated Steam is, that as we increase the pressure the amount of work done by the engine rapidly increases also; but the quantity of heat contained in high pressure steam is very little more than in low pressure. For instance, the units of heat in steam at $110°$C., pressure of $21\frac{1}{4}$ pounds, is $= 640°$C., at $165°\frac{5}{9}$C., or 104 pounds pressure, it is $657°\frac{2}{9}$C., or only $17°$C. more.

Since it is heated by the waste products of combustion passing up the funnel or stack, it is more economical than ordinary steam, but it is by no means economical if this heating is carried to excess. To ensure efficiency it wants little more than drying.

In consequence of its great heat, superheated steam does injury to the internal parts of the engine; it burns the packing, and eats away the cylinders, especially having an injurious effect upon those of indifferent workmanship. As steam is superheated so its elasticity is increased, or the elasticity varies with the temperature. In practice many engineers do little more than dry the steam; for this purpose a small chest, or outer casing, is sometimes fitted round the bottom of the funnel, the steam passes through a short pipe from the boiler to this casing, and is then led away to the cylinder to do its work.

41. Steam when in Contact and not in Contact with the Water.—Remembering the large amount of latent heat in steam, it is evident that the steam contains much more heat than the water from which it is produced. When steam is generated in a boiler and not allowed to escape, as fresh quantities rise from the water the *density* and *elasticity* of the steam must increase; at the same time the boiler is receiving fresh additions of heat. So then, as the temperature increases so do the density and elasticity, which arises from two causes; first, the expansive property of the steam, second, the continual additions of fresh steam keep continually increasing the density and elasticity, while the fire increases the temperature to correspond with the pressure. Steam under such a condition as this is said to be *saturated*, that is, it contains as much vapour as it possibly can for its temperature.

As long as the steam remains in contact with the boiler water, the pressure exhibited by the gauge always corresponds to a certain temperature, and the same temperature in the boiler will always correspond to the same pressure of steam. If the steam be taken from the boiler and further heated in another vessel, in which is no water, we may increase its pressure or elasticity, by increasing the temperature, to almost any extent; but, by doing this, we decrease its density, hence the great distinction between steam when in contact and not in contact with the water is this: when in contact with the water there is a constant ratio between pressure and temperature, but the invariable connection between pressure and temperature does not exist when the steam is not in contact with the water. We may increase the one without augmenting the other.

42. Specific Gravity of Steam.—The specific gravity of steam is its weight compared with an equal volume of air. The specific gravity of steam is ·481, or less than half the weight of the same volume of air.

43. Quantity of Water Required to Condense Steam.—Already it has been shown that the latent heat of steam is nearly five and a half times as much as the sensible heat. Suppose, for instance, we place in an evaporating dish a certain quantity of water, two, three, or four ounces, at a temperature of 22°C., to it we apply the flame of a spirit-lamp, and it boils in $3\frac{1}{4}$ minutes, and continuing the same amount of heat we find the whole is evaporated when 22 minutes 23 seconds more have elapsed. Let us see what we can gather from this:

The water is raised from 22° to 100°, or through 100° − 22° = 78° of heat in $3\frac{1}{4}$ minutes; therefore we conclude that every $3\frac{1}{4}$ minutes heat sufficient to increase the temperature 78° will pass into the water. Into 22 minutes 23 seconds the $3\frac{1}{4}$ minutes will go ($22\frac{23}{60} \div 3\frac{1}{4} =$) 6·887, therefore the total heat passing in after it begins to boil is 6·887 × 78° = 537°·186 C. From this we learn that to evaporate the water takes very much more heat than to boil it, and that the latent heat of steam is 537°·2 C. Again, when this steam returns to water it gives up all its latent heat. We have now to inquire what amount of water is sufficient to reduce this steam immediately to water.

Watt came to the conclusion that $22\frac{1}{4}$ cubic inches of water would condense a cubic foot of steam, or the steam formed from one cubic inch of water, if every atom of water did all it was capable of doing. He reasoned somewhat in this way:—Suppose the latent heat of steam is $537°·2$ C., the temperature of the hot well being $37°$ C., and that of the condensing water $10°$ C., therefore every atom of water will have its temperature raised $37° - 10° = 27°$ C., and as there will be left in the steam converted into water $37°$, the cold water has to take up $637°·2 - 37° = 600°·2$ C.; therefore the total cubic inches of water required for condensation $= \frac{600·2}{27} = 22·23$, or $22\frac{1}{4}$ inches. But as every particle of the water cannot be made to do all its work, he allowed for condensation $28·9$ cubic inches, or a wine pint for each cubic inch evaporated in the boiler, or, in practice, he allowed about one quarter more than was *theoretically* necessary.

EXERCISES CHIEFLY FROM EXAMINATION PAPERS.

1. State clearly the difference between vapour and steam.
2. On what does the boiling point of water depend? State the ordinary temperatures at which fresh and sea water boil respectively. What influences the temperature at which sea water boils.
3. What is high pressure steam? and explain fully the meaning of the expression, "He is using steam of *three atmospheres*."
4. What difference is there between steam in contact and not in contact with the water from which it is generated? What is Mariotte's law?
5. What relation exists between the pressure, density, and temperature of steam?
6. Give the numerical value for the specific gravity of steam. Compare its weight with one or two other gases or fluids.
7. How is steam superheated? What is the difference between common and superheated steam?
8. Of what does sea water consist? Give its constituent parts.
9. What weight of injection water at 80°F. will suffice to condense a given quantity of steam into water at 120°F. (1863)?

Ans. 26·4 for each ounce steam.

Each given quantity of water is raised from 80° to 120° or 40°.
The whole heat in the steam is 967° + 180 = 1147° F.
When this is condensed we leave it at a temperature of 120°.
∴ We have to put in the water 1147 − 88.
As each quantity of water takes in 40°.
∴ Total quantity required is $\frac{1059}{40} = 26\cdot4$.

10. The temperature of the injection water is 70°F., what quantity of injection water will be required to condense a cubic foot of water turned into steam to a temperature of 150°F.?
Ans. 12·86 cubic feet.

11. The hot well is to be kept at a temperature of 125°F., when that of the injection water is 80°F., how much injection water will be required? *Ans.* 23·42 nearly.

12. The temperature of the injection water is 75°F., and the hot well is kept constantly at 127°F.; find the quantity of injection water that is being used. *Ans.* 20·2.

13. Two ounces of water at 60°F. are placed in an evaporating dish, which is covered, except a small opening, by a glass plate. The flame of a gas burner causes the water to boil in $3\frac{1}{3}$ minutes, and the whole is evaporated after 22 minutes more have elapsed. What would you infer as regards the latent heat of steam from this experiment? What is the correct numerical value given by a more exact process (1871)?

In $3\frac{1}{3}$ min. there pass into the water (212 − 60) = 152°F.

1 ,, ,, ,, ,, $\dfrac{152}{3\frac{1}{3}}$

22 ,, ,, ,, ,, $\dfrac{152 \times 22}{3\frac{1}{3}} = 1003\cdot2°\text{F.}$

We, therefore, infer that the latent heat of steam is 1003°; the more correct value is 967°F.

14. The temperature of the injection water is 60°, the steam enters the condenser at a temperature of 212°, the water pumped out of the condenser is at a temperature of 110°; what weight of injection water must be supplied for each pound of steam which enters the condenser (1870)?
(The latent heat of steam at 212° is 966·6).
Ans. 21·37 lbs.

15. What is meant by superheated steam? What advantages are gained by its use (1865)?

CHAPTER III.

RADIATION, OXIDATION, Etc.

The Radiation of Heat—The Absorption of Heat—Reciprocity of Radiation and Absorption—Good and Bad Radiators—Experimental Illustrations—Oxidation of Metals—Effects of Galvanic Action.

44. Radiation.—"Radiation is the transfer or communication of heat from the particles of a heated body to the air or ether." It is a transmission of motion; the vibrations of the heated body, being communicated to the air, sets it in motion. This motion is called *radiant heat*.

45. Absorption.—Absorption is the transmission of motion (radiant heat) from the ether to the particles of any body. Thus when a body is placed in the path of a beam of radiant heat, it partakes of its vibrations, and is set in motion, *i.e.*, it becomes warm, or absorbs the motion.

46. Good and Bad Radiators.—All bodies have not the same powers of radiating and absorbing the motion of heat. Bodies possessing these powers in a comparatively high degree, are said to be *good* radiators and absorbers, and those possessing them in a less degree *bad* radiators or absorbers. Thus earths are good, and water and metals bad radiators. Smooth polished surfaces radiate and absorb much less heat than rough or dirty surfaces. Cylinder covers should be kept perfectly bright, so should teapots and kettles, except the bottoms of the latter, which are required to be good absorbers. A stove for cooking should be bright, while a stove or pipe for heating a room or greenhouse should be rough from the casting mould, and not painted. Radiation can be prevented by clothing with non-conductors. Cylinders have a wooden casing made for them; boilers are built into brickwork with layers of ashes, sawdust, etc., around them; and steam pipes are covered with matting, etc., to prevent radiation.

47. Reciprocity of Radiation and Absorption. — It is interesting to note the reciprocity which exists between the power of a body to communicate the motion of heat to the ether, *i.e., to radiate*, and its power to receive motion from a heated body through the medium of the ether, or to absorb. In other words, *good radiators are good absorbers*, and *vice versa*. Thus the earth (rocks, etc.) quickly absorbs the radiant heat of the sun, but no sooner does the sun set than the heat radiates from it; while the sea slowly gets warm, and retains its heat much longer than the earth.

48. Experimental Illustrations — *Radiation.*—Place a common mercurial or other thermometer about a yard distant from the fire, or any other heated body, and instantly it will indicate an increase of temperature, the motion having been communicated by the fire to the air, and by *absorption* it has been accepted or absorbed by the glass of the thermometer, and by conduction transmitted to the mercury. The melting of ice when placed before the fire, warming your hands, the sun heating the rocks, and they in their turn warming the air, are all instances of radiation from the heating body and absorption in the body warmed.

Good and Bad Radiators.—Take two hollow vessels, one of metal and one of earthenware, and fill them with boiling water. If a thermometer be held outside and close to the metal one, it will be found to show a much less increase of heat than when placed outside and close to the earthenware vessel, and the water in the earthenware vessel will cool much more rapidly than that in the metal one. As a rule metals are bad radiators, and earths good radiators. A slate held before a fire will receive or absorb much more heat than a brass plate, and so on. Dry atmospheric air will absorb no heat whatever; in fact, the rays of the sun, or radiant heat from artificial sources, may pass through it without altering its temperature. On high mountains the direct rays of the sun may be almost unbearably scorching, while the air is perfectly cold, and the traveller has only to withdraw into the shade to feel the freezing chill of the atmosphere. For the same reason, that the air is perfectly dry, ice is often formed at night in the desert of Sahara, where, during the day, the direct rays of the sun make it a fiery furnace.

Reciprocity of Radiation and Absorption.—If when a red hot ball be placed between two plates of pewter and glass respectively, and the different temperatures of the two plates and of the air immediately around them be noticed, it will be seen that the glass not only receives or absorbs much more heat than the pewter, but also that the radiation from the glass is much more powerful than from the pewter.

From the remarks accompanying the definition of reciprocity of radiation and absorption, it will be seen that where radiation is powerful we may expect good absorption, and *vice versa*.

49. Galvanic Action and Oxidation of Metals.—Metals are subject to two kinds of deterioration—galvanic action and oxidation. When two different metals come in contact, especially if they are constantly wet, a galvanic current is induced which results in the decomposition of one of the metals, or one destroys the other. For instance, who has not observed that old iron railings are frequently wasted away towards the bottom, close against the lead that fastens them into the stone? The reason is, that a galvanic current passes from one to the other, and the soft lead wastes away the hard iron. If we take, in the following order, silver, copper, tin, lead, iron, and zinc, we have these in their relative positions as regards galvanic action, and the farther they are from one another in this list the greater the effects of galvanic action. Those coming first in order will destroy any that follow them. Copper, when in contact with tin, lead, iron, zinc, etc., will waste them away, but not silver—the silver will eat away the copper, tin, lead, etc. When copper pipes are fastened by iron bolts or screws, the iron is soon destroyed, especially in damp situations.

Oxidation is a chemical action. When iron rusts we have an instance of oxidation. The oxygen of the air enters into chemical combination with the iron, and forms oxide of iron or rust. When the oxygen of the air combines with the copper, we have oxide of copper or verdigris. Zinc in the same way becomes covered with a layer of the oxide of that metal when exposed to the air. Thus it is necessary that metallic substances should be covered with paint, grease, etc., when exposed to the air, or otherwise oxidation may proceed with sufficient rapidity to injure them.

Two other facts, which are closely allied to oxidation and galvanic action, may be stated, namely :—when superheated steam is employed in jacketed cylinders, and much tallow introduced, it is found that the tallow is decomposed, and *carbonises* the piston, so that it becomes more like a piece of plumbago than anything else. Cast iron long immersed in sea water may be cut with a knife.

CHAPTER IV.

THE ENGINE BEFORE WATT, AND WATT'S ENGINE AND IMPROVEMENTS.

Savary's Engine—Newcomen's Atmospheric Pumping Engine—Its Defects—The Discoveries of Watt—The Separate Condenser—The Expansive Working of Steam—Its Economy—Its Value in Regulating the Power of an Engine—Details connected with Watt's Single Acting Pumping Engine—The Steam Cylinder—Valves connected with Cylinder and their Action—The Condenser—The Air Pump—The Foot Valve—The Delivery Valve—The Snifting Valve—The Hot Well—The Piston-Rod—Connecting Rod and Crank—Stuffing Boxes and Glands—Parallel Motion—Method of Starting the Engine and of Regulating its Speed by the Governor—The Throttle Valve—The Cataract—Eccentric.

50. Savary's Engine.—Savary's was the first steam engine employed to pump water. He took out his patent in 1698. His engine consisted of a cylinder, in which steam was employed to produce a vacuum only, after which he relied upon the pressure of the atmosphere to raise the water. At the top of his cylinder were two openings, each fitted with a pipe and a stop-cock. These were so arranged that the same handle opened one stop-cock and shut the other simultaneously. One pipe communicated with a boiler and admitted steam to the cylinder, the other with a cistern of cold water. From the bottom of the cylinder a pipe led down to the water. It acted thus: Suppose the handle of the stop-cock moved, and steam admitted to the cylinder, directly it was filled the handle was pushed back, and a dash of water from the other cock condensed the steam and formed a vacuum; then the pressure of the air on the water at the bottom of the mine forced the water up into the cylinder, which was prevented from returning by a valve opening upwards; on a second admission of steam, its elastic force acting on the water drove it through a valve in the side of the cylinder opening outwards; this steam was again condensed as before, etc. We thus see the principle upon which it acted. The water was first forced by atmospheric pressure into a vacuum, after which the elasticity of the steam, pressing upon its surface, was made to raise it still higher through another passage. The inefficiency of this machine is apparent. Its defects were: that steam was used in a cold cylinder; that the steam was always in contact with cold water, and, therefore, the greater part of it was lost; that the engine was limited in its range and purpose; that it must be always far down in the mine from which the water was raised.

51. Newcomen's Engine.—Thomas Newcomen was a Devonshire man, and the first to work out the idea of a piston (at least in England). His engine was used for pumping. In fact, the one idea of the early labourers at the steam engine was to adapt it, or to invent a machine, to pump water out of the Cornish mines.

Newcomen placed his cylinder immediately above his boiler, from which steam passed directly through a stopcock. As soon as the piston was at the top of its stroke, a cock was opened and cold water admitted into the cylinder to condense the steam; a vacuum being thus obtained, the pressure of the air, 15 lbs. on the square inch, immediately drove down the piston, which was attached by a chain to the end of a sway beam moving on its centre. The piston being thus forced down by atmospheric pressure pulled up the other end of the beam at the same time, and with it the pump rods, water, etc. When fresh steam was admitted it forced up the piston against the atmosphere, while the weight of the pump rods, etc., at the other end, assisted the steam. The weight of the pump rods, etc., was generally made equal to half the pressure of the air on the piston. This engine raised 7 or 8 lbs. for each square inch of the piston. Newcomen's was a *single acting* engine, because the steam acted on one side of the piston only.

In Newcomen's engine, as represented in the accompanying figure, A P is the ashpit, F P the fireplace, B the boiler, S C a stop-cock to admit the steam into the cylinder H from the boiler B. The cylinder was bored as truly as possible, open at the top and closed at the bottom, being connected with the boiler by a short pipe containing the steam-cock. A piston *p* was made to move up and down in the cylinder, as air-tight as practicable, by packing its edges with hemp and covering the upper surface with water. The piston-rod *r* was attached by a chain *c* to the circular arc *c d*, forming the end of the beam *e* C *d*, *which was now for the first time introduced*. The beam worked on its centre C, and was formed of strong

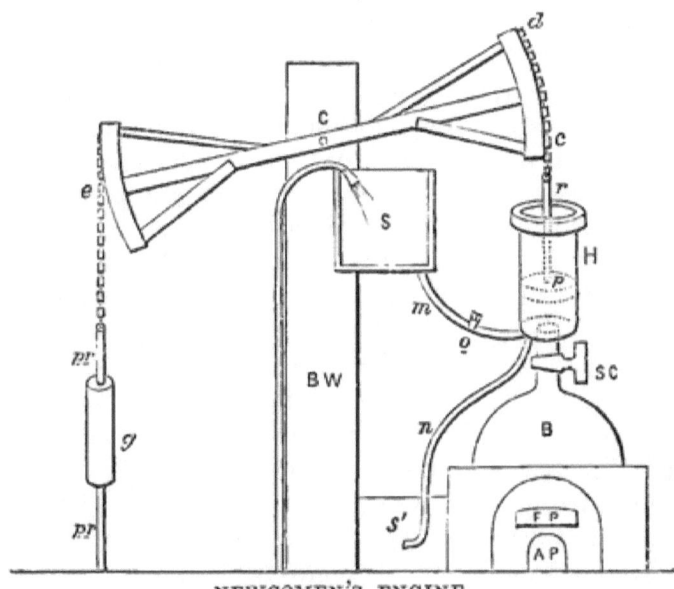

NEWCOMEN'S ENGINE.

timbers firmly put together and strengthened by iron bars and straps. The whole beam was supported on a strong brick wall, B W. To the chain *e* attached to the other arc was fastened the rod $p\,r$ of the pump to be worked in the mine. The power of the engine was in the down stroke. The pump rod was made heavy enough to act as a counterpoise by attaching weights g to it, so that it was heavier than the piston, piston-rod, friction, etc. When the cock S C was opened and *air* admitted, it would rise freely without violently jerking out the piston p. A safety valve was placed on the top of the boiler. The manner in which the engine worked may be thus described :—

The boiler B was filled with a proper quantity of water, and the steam "got up" to a pressure a little above that of the atmosphere. The cock S C was opened (supposing the piston at the bottom of the cylinder), and the steam entered the cylinder, when the piston ascended partly through the force of the steam, but chiefly in obedience to the counterpoise weights g. Just before the piston reached the top of the cylinder, the steam-cock was shut and another cock o opened, which allowed water from the cistern S to flow through the pipe m and condense the steam in the cylinder, producing a vacuum, when the pressure of the external air, acting on the top of the piston, caused it to descend with a force proportionate to its area; and as this force amounts to nearly 15 lbs. on the superficial inch, it was fully competent to raise the end of the beam e, and with it the pump rods and water. We thus see that the real work was done by the *atmosphere*, and why it was called an atmospheric engine. All the 15 lbs. was not effective.

Originally, it was much less perfect than here described, for the condensation was in the first instance performed from the outside of the cylinder. The admission of water into the cylinder to condense the steam was discovered accidentally, through some holes being in the piston of an engine which permitted the water, placed upon it to keep it air-tight, to run through and condense the steam, although we must remember Savary had introduced steam into his cylinder and *condensed it in the* *cylinder*. The great difficulty of opening the cocks at the proper moment was conquered by Humphrey Potter,* who attached some strings and catches to the cocks of an engine he was employed to work at Wolverhampton, in order to release himself from the trouble of attending them; his contrivance gave the first idea of "hand gear." The greatest nicety and attention on the part of the workman was necessary in turning the two cocks at the proper moment; for if steam were permitted to enter the cylinder for too great a length of time, the piston would be carried out of it or blown out of its place; while, on the contrary, if not opened soon enough, it would strike against the bottom with sufficient force to break the cylinder. The steam was liable to become mixed with air which was disengaged from the injection water. This air, together with the injection water, was discharged by a pipe n into the cistern s'. The pipe n terminated in a valve to preserve the vacuum, which valve, from the peculiar noise it made, was called the *snifting valve* or *snifting clack*.

52. **Defects of Newcomen's Atmospheric Engine.**—We have already hinted that it was named an *atmospheric engine*, because it depended upon the pressure of the atmosphere to perform the down stroke, or to do the real work. Its great defect was that the steam was used in a cold cylinder and condensed in a hot one; *i.e.*, it was cold when required to be warm, and warm when it should be cold. It has been estimated that, by condensing the steam in the cylinder, three-fourths of the power of the engine were lost.

53. **The Discoveries of Watt, and Separate Condenser.**—Watt, having the model of an atmospheric engine, such as we have just described, to repair, asked himself the question, whether it were not possible to prevent the wasteful expenditure of steam. He saw intuitively the great defect of the engine, and set himself to solve the problem of a separate condenser. In this he completely

* Millington's *Mechanical Philosophy*.

succeeded, and never left the steam engine until it was comparatively a perfect machine. The annexed figure is a fair representation of the *great* improvements he introduced.

A B is a large casting, within which is placed the condenser C, the air pump A P, and the hot well H W. V is the piston or bucket of the air pump, with its two valves shut down, but shown by dotted lines as they will appear when the piston V is descending. E P is the

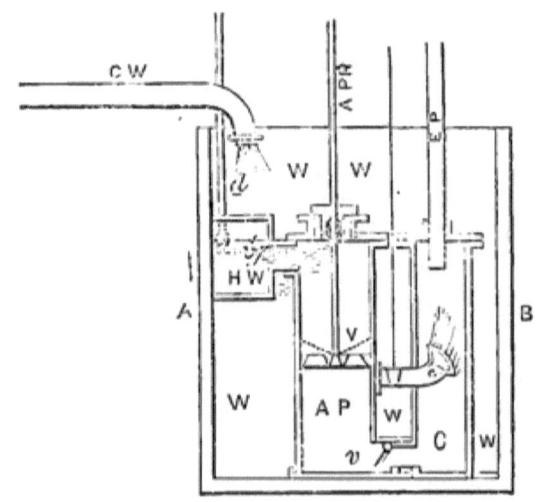

CONDENSER AND AIR PUMP.

exhaust pipe, to convey the used steam from the cylinder into the condenser C. C W is a pipe bringing cold water from the pump, *v* the foot valve, *v'* the delivery valve. W W W W is water surrounding the condenser and air pump, to keep the condenser cold.

Let us suppose that the steam, having been used, comes from the cylinder through the exhaust pipe E P. The moment it enters the condenser, it is met by a scattered jet of cold water from the rose head *c*, and is condensed. The condensed steam and water fall to the bottom of the condenser, and pass or are drawn through the foot valve *v*.

Then the piston or bucket V of the air pump comes down into the water; the pressure of water opens the two butterfly valves, and the water passes through the valves, and so gets above the piston. When the piston is drawn up, the two valves are closed by the weight of the water above them, which is next forced or delivered into the hot well H W, through the delivery valve v', from whence a portion of it is pumped into the boiler through d, a part of the *feed pump*. As the air pump ascends a vacuum is formed in A P, at least as good a vacuum as exists in the condenser C, so that the condensing water passes by gravity, etc., through the foot valve v, or "follows the bucket." As the air pump descends, we see v must close, so must v'; on the contrary, as it ascends, both delivery and foot valve will open.

All water contains air more or less. The heat of the steam disengages the air from the condensing water, which would rise through the exhaust pipe, and prevent the proper escape of steam, and counteract its pressure if not got rid of.

The air pump was, therefore, added by Watt to his invention of the condenser, to prevent air from accumulating and obstructing the engine.

54. The Expansive Working of Steam—its Economy —its Value in Regulating the Power of an Engine.— By expansive working of steam is simply meant this: that the steam from the boiler is admitted to the cylinder during only a part of the stroke; this admission being stopped, the steam in the cylinder has then to complete the remainder of the stroke by its expansive property. By Boyle's or Mariotte's law, the pressure of steam varies inversely as the space it occupies. For instance, suppose we have a cylinder full of steam at a pressure of 30 lbs. on the square inch, if we compress it into one half the space, the pressure will then be 60 lbs. on the square inch; while, if we allow it to fill another cylinder of the same size, as well as the one it originally occupied, or, in other words, allow it to double its volume, its pressure will be only 15 lbs.; *i.e.*, compress it into half the space its pressure is doubled, allow it to occupy twice the space the pressure is only one half.

This figure will give a fair idea of the expansive working of steam, and its economy and value in regulating the power of an engine. Suppose steam of 60 lbs. is admitted to a cylinder 6 ft. long, and cut off at ⅓ the stroke, then

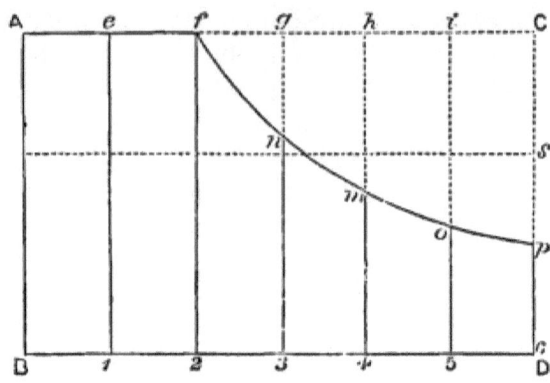

AN EXPANSION DIAGRAM (THEORETICAL.)

during the first and second feet the pressure is 60 lbs.; at the end of the third foot, the steam will occupy one and a half (3/2) the space, the pressure at the end is therefore ⅔ of 60 lbs. = 40 lbs.; at the end of the fourth foot the steam fills twice the space, the pressure is then ½ of 60 lbs. = 30 lbs. So that from the end of the second foot to the end of the fourth the pressure has fallen from 60 lbs. to 30 lbs. At end of the fifth foot the steam fills two and a half (5/2) times the space, hence pressure is ⅖ of 60 lbs. = 24 lbs. Hence we have steam being admitted at 60 lbs., and

At end of first foot its pressure is 60 lbs.
,, second ,, ,, 60 lbs.
,, third ,, (⅔ of 60) = 40 lbs.
,, fourth ,, (½ of 60) = 30 lbs.
,, fifth ,, (⅖ of 60) = 24 lbs.
,, sixth ,, (⅓ of 60) = 20 lbs.

We may say that steam of 20 lbs., by giving it a great initial pressure, and cutting it off at one third stroke, has been made to do work equal to steam whose average pressure is above 39 lbs. Again, during the first and second feet the pressure of steam was 60 + 60 = 120 lbs., during the remainder of the stroke the pressure = 40 + 30 + 24 + 20 = 114 lbs., so that without any additional expenditure of steam, by merely allowing it to expand, we get almost double the work out of it when the steam is cut off at $\frac{1}{3}$ stroke. We nearly double the work of one-third of a cylinder of steam by mere expansion. This sufficiently shows its superior economy. If in our figure we cut off from g $\frac{2}{3}$ of its length, from h $\frac{1}{2}$, from i $\frac{2}{5}$, from C D $\frac{1}{3}$, and through the points draw the curve f n m o p, then that curve will represent the gradually decreasing pressure of steam. It is an hyperbolic curve. We may partly see the reason for the rule to find the units of work done by a piston in one stroke, which is

$$q\,p + q\,p\,\log_{\varepsilon} \frac{l}{q}$$

where q is the portion of the stroke moved through before steam is cut off, l is the length of the stroke, and p the pressure at which steam is admitted. For a proof of this formula the reader is referred to the volume on *Steam* in the Advanced Series.

55. Watt's Single Acting Engine.—In this engine A B is the cylinder, P the piston, P R the piston rod, S the steam pipe, D leads to the exhaust, a b c are three valves on one spindle, a the steam valve, b the equilibrium, and c the exhaust.

The following is an explanation of the action of this engine:—Steam comes along the steam pipe S from the boiler, when the valves a b c, being in the position shown in the figure, with a and c open and b closed, the steam enters the cylinder A B in the direction marked by the arrows with tails, and drives the piston down, causing the pump valves at the other end to ascend. Steam that may have been under the piston in E can freely pass away to the exhaust D. The moment the piston is at the bottom of its stroke, the valves move to their second position, so that a and c rest on their seats o, while b is opened. Then the steam that drove the piston down can run through valve b, in the direction shown by the arrows without tails, get under the piston P, and assist in driving it up. The pump rods at the other end are balanced by a counterweight to assist this expanding steam. The action is then continuously repeated: a and c open, steam enters through a, drives down P, and the steam under P escapes through c, then a and c are closed, and steam runs round through b to assist the upward motion of the piston.

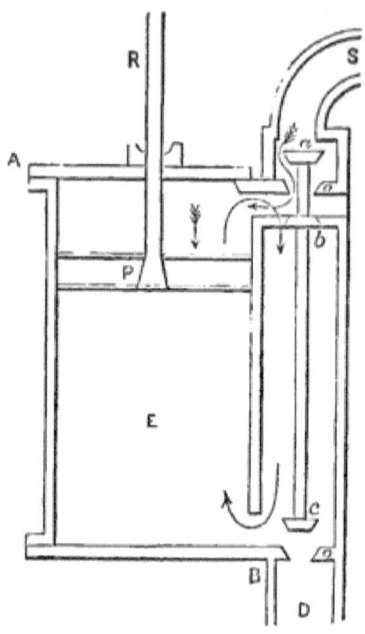

56. Snifting Valve or Snifting Clack was so called from the noise of its action; in Newcomen's engine, page 47, at s', was a valve for the escape of the condensing water and air, through which they escaped from the cylinder with a snifting noise. In Watt's engine a snifting valve was also fitted which communicated with the bottom right hand corner of the condenser, so that the air might escape, but it was chiefly fitted to assist in blowing through previously to starting the engine; the steam admitted by the blow-through valve escaped by the snifting or tail valve, as it is called. A blow valve is not always fitted now, because steam of a much higher pressure is used. Hence neither blow valve or snifting valve are so important as formerly.

57. Double Acting Engines.—When steam drives the piston both up and down, the engine is termed double acting. All our modern engines are double acting; but

Newcomen's was an atmospheric and single acting engine, the piston being driven up by steam, but down by atmospheric pressure. Watt's first engine was single acting; the steam drove the piston down, while the weight of the rods, etc., at the other end of the beam, brought it up.

58. Clearance.—When a piston makes its stroke it is not allowed to touch the top and bottom of the cylinder for fear of knocking them off.

The space between the top and bottom of the cylinder and the piston, when the latter is at the end of its stroke, is the clearance.

59. Cushioning.—When the steam is shut in before the end of the stroke, the piston acts against it as against a cushion, and so is brought gradually (comparatively speaking) to rest. Suppose the piston is in the position A B when the steam is shut in, and that from A to C is 12 inches. Let us also suppose that the elastic force of the steam remaining behind in A F is 2 lbs., when the piston gets to D, 6 inches down, by Mariotte's law its elastic force will be 4 lbs.; when at E, 9 inches down, it will be 8 lbs., etc. So we see at once the effects and advantages of cushioning, and that it must bring the piston gradually to rest by destroying its momentum.

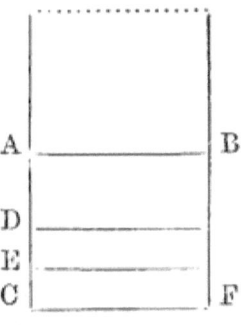

60. The Piston and how Fitted—Packing, etc.—As the piston is a most important part of the engine, great care and thought have been bestowed upon it. It must be perfectly steam tight, and, at the same time, it is required to move easily within the cylinder. A cylindrical piece of iron is chosen and turned about a quarter of an inch smaller in the diameter than the bore of the cylinder, and around it is cut a deep groove square in section; into this is fitted a metallic ring of brass or steel, but generally cast iron; this ring either fits steam tight against the cylinder by its own elasticity, or is forced against it by springs or compressed air. Formerly "packing" was much used, when some rope yarn was platted the exact size of the square groove, the precise length was cut off, and the ends neatly sewn together—care being taken that no turns were left in the yarn. The whole was well greased before it was fitted in. Metallic piston rings are now most in fashion, the piston being composed of two distinct parts—the piston proper and the junk ring. The junk ring is bolted on to the piston by bolts tapped into the piston and heads recessed into the junk ring. A metal ring is next turned exactly the size of the cylinder, and then cut; when cut, we know such a ring will develop its elasticity, and some force will be required to place the ends in contact again. It thus forms a powerful spring, and is placed between the junk ring and the piston, where a place has been left for it. The piston is now complete, and the spring or metal ring, being compressed into its proper position, the whole is placed within the cylinder, forming a very steam tight easy piston.

Pistons are seldom packed now, but the air pump bucket is; because packing is cheaper, and also because in this case it answers better, for a large amount of galvanic action sets in and eats away the metallic parts of the piston of the air pump.

61. **Stuffing Boxes and Glands.**—These are used in several parts of an engine. A good example may be seen in the fig. in page 69. The piston-rod enters the cylinder through the stuffing box $s\ b$; while the packing, the part marked so dark within the stuffing box, is pressed down in its place by the gland $g\ d$; bolts pass through the flanges of both, so that when the steam leaks through the cover by the side of the piston-rod, we have only to screw the gland down on to the packing, and the leak is stopped by the packing, being forced against the piston-rod. A depression will be seen round the top of the gland close to the piston-rod, it is to hold oil or tallow to lubricate the piston-rod.

62. **Method of Starting the Engine.**—The engine is

started by opening the stop valve, which allows the steam to run from the boiler to the cylinder; when the steam is in the cylinder it gives the required reciprocating motion to the piston; the supply is regulated by proper openings and valves, to be afterwards explained. In starting an engine, where the single eccentric is fitted, we must notice which way the engine has to go; then with the starting bar move the slide up or down, according to the direction we wish to send the crank, when the piston, etc., will begin to move; a stop on the shaft carries the eccentric with it; when in the proper position, the end is dropped on to a stud or pin and fixed in its place, and the engine itself works the slides. In explaining Stephenson's Link Motion, an effort will be made to render this part of the subject a little clearer.

63. **The Parallel Motion.**—Although the parallel motion has been almost superseded by simpler pieces of mechanism, such as guides, quite as efficient, yet a description cannot be wholly omitted.

If the end of the piston-rod g had been connected to the end of the beam, the piston-rod would have been bent alternately to right and left as the beam rose and fell, and a continual jarring would be going on, constantly destroying the stuffing box, and rendering the cylinder leaky.

Let us suppose that the simple lines in the adjoining figure represent the parallel motion, C h is half the beam, $h\,g$ is the main link, $c\,d$ the radius bar or bridle rod. As h moves up and down, it describes an arc of a circle with its convexity to the left. Now $c\,d$, the radius bar, moves on its fixed centre c, consequently the point d will describe an arc with convexity to the right; so h throws $g\,h$ to the left, and $c\,d$ throws $d\,e$, and with it $g\,h$, to the right. Therefore it is evident that if these links and rod be proportionately adjusted, we shall have an arrangement that will compel the point g, and with it the whole piston-rod, to move exactly perpendicularly. To accomplish this there are joints at g and d.

58 STEAM.

To find the proper length of the bridle rod,

Divide C h in e, so that
$$C e : c d :: d o : o e$$
where o is the point to which the air pump rod is attached,

$g\,d$ or $h\,e : C e : d o : o e$
$\therefore h e : C e :: C e : c d$
$$\therefore c d = \frac{C e^2}{h e}$$

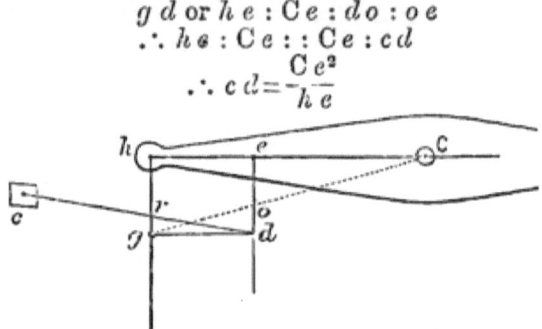

PARALLEL MOTION.

The parallel motion will work most accurately when the radius rod, from c to d, is about the same length as the beam from C to h, they should therefore be kept as nearly equal as circumstances will permit.

64. The Governor.—The governor consists of two balls, A and B, fixed on the ends of two arms, and so arranged that they can freely revolve round the spindle C D. Motion is imparted to the balls either by a pulley, which is driven by a cord passing over another pulley on the main shaft by the side of the fly wheel, or else by a pair of bevel wheels placed immediately below D.

When at rest, the balls will remain close to the governor spindle, as in the figure, but when in motion the faster it moves the farther the balls will fly asunder by centrifugal force. As they separate, the arms A C and B C will extend outwards, and will bring up with them the short arms G H and E F, which will move up the collars I, L, when the arm M N will pull point N to the left; P is a fixed joint and P Q is firmly attached to P N, so that point Q will be lifted up and close the throttle valve V in the steam pipe S, by means of two arms, one of which, Q V, is shown in figure, moving the valve on its spindle. Thus, the faster or slower the main shaft moves, the faster or slower will the governor move, and close or open the throttle valve and regulate the supply of steam, so that the engine may always be moving at the same velocity.

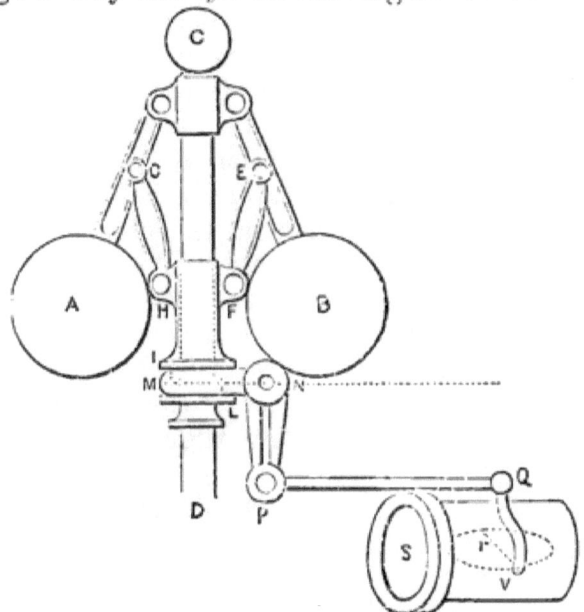

GOVERNOR AND THROTTLE VALVE.

In flying outwards, the balls attain a certain vertical height, which height, and the length of the pendulum, to vibrate in a given time, are calculated, as shown in the volume on *Steam*, in the Advanced Course of this Series. The weight of the balls does not affect the action of the governor at all, for if a heavy ball increases the centripetal force, it also increases the centrifugal in the same ratio. It is called the conical pendulum or pendulum governor, because its motions are regulated by the same laws as those which regulate the ordinary pendulum.

65. Throttle Valve.—From the last figure a good idea can be obtained of the throttle valve. It is a circular or elliptical plate moving on a spindle. Its opening, as regulated by the governor, determines the volume of steam that shall pass to the cylinder.

66. The Cataract.—The cataract supplies the place of the governor in the single acting Cornish pumping engines. It consists of a small pump plunger a and barrel $b\ c$ set in a cistern of cold water A B; d is a valve opening inwards, so that when the plunger a ascends, the water passes through d from A B into $b\ c$; f is a cock opened and shut by the plug e, moved by the plug rod g, worked by the beam overhead. If the plunger be forced down, the water will pass through f in proportion to the opening of f. When the beam has moved fully up, it liberates the rod that works the plunger; then as the chamber fills with water through d as the plunger ascends, so when the latter comes down the pressure of water will close d, and the weight of the plunger will force the water through f as rapidly as the opening will allow. The way it is carried away is not shown in the figure. If the cock be shut, the plunger cannot descend; if only slightly opened, it will descend gradually, etc. As soon as a certain quantity of water has passed through f, its weight opens the injection

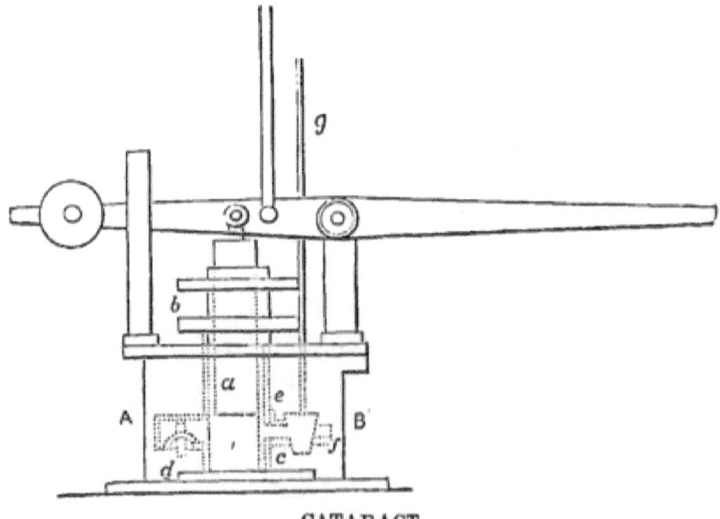

CATARACT.

valve, and condensation takes place, when the engine can complete its stroke; for the engine can only make its stroke as the water is supplied for condensation. It thus regulates the speed of the engine; for if the cock be fully open, condensation takes place at once, and if only partly open, condensation will be delayed till the water is supplied.

67. Marine Governor.—Owing to the unsteady motion of a ship, arising from pitching, rolling, etc., the ordinary pendulum governors are unfitted to regulate the speed of the engines. Mr. Silver has solved the problem how to adapt a governor to a marine engine. He has employed several arrangements for carrying out his ideas. The one of which a section is here shown seems the best adapted to the purpose.

A B is a small fly wheel about 18 inches in diameter, on which are fixed two fliers or vanes, F. The faster the engine goes, the greater resistance will these vanes offer to the air. P is a pulley worked by a cord and fixed on the spindle $s\,s$, while E is an eccentric and K a lever. To E

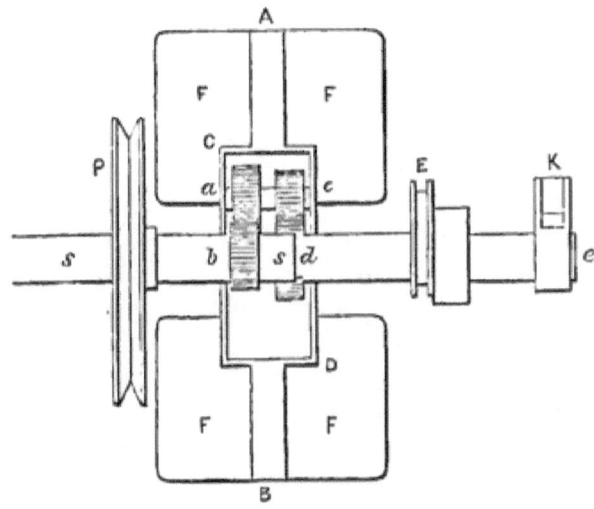

SILVER'S MARINE GOVERNOR.

at the top of the pulley, for the position given in the figure, is affixed a spring. The engineer has to tighten up or slacken this spring according to the speed at which it is intended to drive the engines. K is the lever from which the motion is conveyed to open or close the throttle valve. Within C D are four pinions to communicate the action necessary to affect the purpose of the contrivance. Sometimes there are six pinions, one below b and d respectively.

At the uniform speed of the engine, it revolves together in connection with the engine as the motive power; but when accelerated by the running of the engine, as when the screw is out of the water, the increased pressure on the governor fans, or blades, causes the motion to act on the eccentric E, and the lever K carried on the tube $d\ e$. (We must understand $d\ e$ is not a continuation of $s\ s$). Then the spring, attached to E or the arm to K, according to whichever arrangement is adopted, acts to close the throttle valve. The pinion b, keyed on the solid shaft $s\ s$, gearing in the wheel a, which runs on a loose pin $a\ c$, transmits the motion to c and to d, a pinion keyed on the tube $d\ e$, which acts upon the lever, regulates the speed of the engine. It is excessively sensitive, and the least increase or retardation of speed causes it to act upon the valve. When the pulley is running very fast, the inertia of the fliers and the resistance of the air will not allow the fliers to go as fast as the pulley, so the pinion a runs as it were back on b (or b overtakes a), and acting on the spring at E and the lever at K, the latter closes the throttle valve. In one arrangement of this governor, the spring itself works the valve.

68. **To Close the Throttle Valve.**—To maintain the spring at the elasticity at which it is set requires a certain speed, and when the engine falls below this speed, the spring slackens itself, and allows the valve to open.

69. **Eccentric.**—The eccentric consists of a disc of metal encircled by a hoop or strap, to which is attached the eccentric rod; in the disc is a hole to pass it on to the main shaft. The centre of the eccentric does not coincide with the centre of the shaft. When the shaft revolves it carries with it the disc, which, moving within the hoop, gives a reciprocating motion to the eccentric rod.

A B is the eccentric, B C the eccentric rod. *a b c* is the solid disc that can move round within the strap or

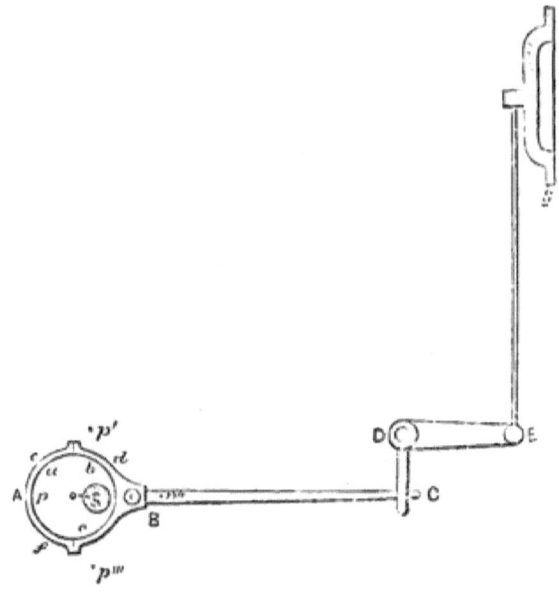

ECCENTRIC, ECCENTRIC ROD AND GEAR.

band *d e f;* *o* is the centre of the disc. S is the main shaft, on which the disc is tightly keyed. As the eccentric or disc revolves within the strap, it will be easily seen that the point *p*, moving round, will come into the positions *p′ p″* and *p‴*, and that the point C will be thrown alternately to the right and left. C D E is a bell-crank lever supported on D, a fixed point, and therefore since C moves alternately right and left, E moving along the arc of a circle will give a vertical reciprocating motion, and alternately pull the slide *s* up and down. The distance between the two centres *o* and S (marked by a line in the figure), is called the *throw* of the eccentric. The disc is generally keyed on one-sixteenth of a revolution in advance of being at right angles to the crank.

EXERCISES CHIEFLY FROM EXAMINATION PAPERS.

1. Give a description of the steam engine in use before the time of Watt, with an account of his improvements (1863).

2. Give a sketch of a blow valve and a snifting valve, and show why these valves require no springs nor weights to keep them in their seats (1863, 1864).

3. Mention the distinguishing features of the atmospheric single acting and double acting engines, what kind of engine is generally fitted to steam vessels, and what kind is best suited for land carriage (1864).

4. The total pressure on a pair of equal pistons is 90 tons, at the rate of 45 lbs. on each square inch, find their diameter (1866).
Ans. 53·4 inches.

5. Give an account of the steam engine in use before the time of Watt with an account of his improvements (1866).

6. Explain the way in which the eccentrics of marine engines are fixed on the shaft. Explain also the method of obtaining the back motion (1866).

7. The area of a piston is 4876·84 square inches, find the diameter of the air pump, which is half that of the cylinder; find also the capacity of the pump, supposing it similar to the cylinder (1867). *Ans.* 39·399 and 1219·21.

8. The area of a piston is 4476 square inches, and the diameter of the piston-rod is $\frac{1}{8}$th that of the piston, find it (1868).
Ans. 9·43.

9. What is the foot valve? Is it a necessary appendage to a steam engine. If it is not used, what arrangements must be made in consequence (1867)?

10. The pressure of steam is 15 lbs. on the square inch, and that of the uncondensed vapour is 2 lbs. Compare the effective force in the up and down stroke respectively (1868).
Ans. 16:15.

11. Describe generally the improvements introduced by Watt into the steam engine (1868).

12. What are the foot valve and delivery valve? What is meant by blowing through? How is it effected (1868)?

13. Describe Newcomen's atmospheric pumping engine, and point out its defects (1869).

14. Explain the manner in which the steam acts in Watt's single acting pumping engine. Why is this engine so much more economical in steam than the old atmospheric engine (1870)?

15. Why is it economical to cut off the steam before the piston has gone to the end of the cylinder? The length of the stroke of an engine is 8 feet, the pressure of the steam on entering the cylinder is 30 lbs. on the inch; at what point should the steam be cut off so that the pressure at the end of the stroke may be 5 lbs. per inch (1870)? *Ans.* $\frac{1}{6}$.

16. Describe the eccentric for working the slide valve of a steam engine. How is it thrown in and out of gear? How is it attached to the slide rod in an oscillating engine (1870)?

17. In what manner is the work done by steam estimated? What is the numerical expression for the work done when steam, at an effective pressure of 20 lbs. on the square inch, forces a piston 20 inches in diameter through a space of two feet against a resistance (1871)? *Ans.* 12566·4 units.

18. It was stated by Watt that neither water nor any other substance colder than steam should be allowed to enter or touch the steam cylinder during the working of an engine. Show that this rule was not adopted in the case of the atmospheric engine, and describe the arrangements by which Watt gave effect to it (1871)?

19. What is done by the air pump in a steam engine? What are the foot and delivery valves, and where are they placed? Describe some gauge for estimating the exact pressure of the air, or uncondensed vapour in the condenser (1871)?

20. Explain the action of the governor and throttle valve in regulating the speed of an engine (1869).

21. Describe the arrangement of the condenser and air pump of a condensing engine, and the valves connected therewith.

E

CHAPTER V.

BEAM ENGINE AND DETAILS.

Double Acting Condensing Beam Engine—Principle upon which it Works, etc.—Details of the Various Parts—Cylinder—How Constructed—Ports or Openings into the Cylinder, etc.—The Form of Slide Valve in Common Use—The Locomotive or Three-Ported Valve—The Lap on a Valve—The Eccentric—The Lead of a Valve—Cushioning the Steam—Clearance—Details of the Piston—Metallic Packing-Rings—The Expansion Valve and the Gear connected with it—The Supply of Water for Condensation—Blowing-through—Gauges for the Condenser—The Barometer Gauge—Method of Estimating Pressure by it—Errors in this Method, and correction of the Same—The Fly Wheel—The Principle of an Equilibrium Valve—The Double Beat Valve—The Crown Valve—The Throttle Valve—The Gridiron Valve—The High Pressure Engine without Condensation—The Expansive Principle as Applied in the Double Cylinder Condensing Engine.

70. Definition.—A *double acting engine* is one in which the piston is driven both up and down, or backwards and forwards, by the action of the steam. A *condensing engine* is one in which the steam, after it has driven the piston up or down, is led away to a separate place, where it is condensed by the application of cold water to it. A *non-condensing* engine is one in which the steam, after it has driven the piston up or down, is allowed to escape into the air, and is not condensed, as the locomotive. Non-condensing engines are misnamed high pressure engines, while condensing engines are erroneously termed "low pressure" engines. Non-condensing or low pressure engines were so named because Watt at first used steam but little above atmospheric pressure, and in some cases even below 15 lbs; while, on the contrary, when steam began to be used without condensation, being necessarily much above the pressure of the atmosphere, such engines were called "high pressure." But now steam of as great a pressure is used in condensing as in con-condensing. Hence it is better to divide engines into the two classes, (1) condensing, (2) non-condensing.

71. Beam Engines.—Newcomen's was a beam engine and so was Watt's, but the latter was far more perfect than the former. The crank was not patented in time by Watt, he therefore used the sun and planet wheel for a crank. The beam was so advantageous and so thoroughly incorporated in the steam engine, that to early engineers it seemed an inseparable part of it as much as the cylinder and piston, therefore when it came to be adapted to marine propulsion, the side lever was the only modification that presented itself. The great advantage of the beam engine is, that to the parts requiring it, it gives a longer leverage, and therefore greater power; a long connecting rod is employed, and thus an immense advantage is gained. Again, a fly wheel was used with it to accumulate power.

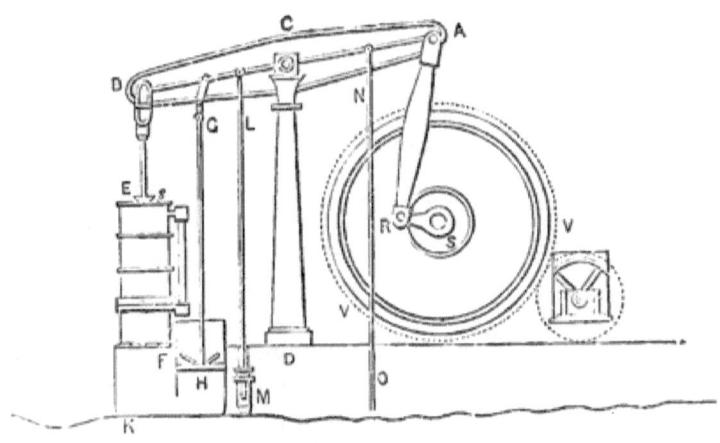

BEAM ENGINE.

A B is the *beam* moving on its main centre C, supported by a frame and pillars, of which C D is a front one; B E is the *piston-rod* working in and out of the *stuffing box s*, at the top of the *cylinder* E F; G H is the *air pump rod;* H the *air pump*, with the *condenser* H K (only part of which is shown); L M is the *feed pump rod;* M the *feed pump*, into which the plunger is seen descending; N O is *the pump* to force up water for condensation; A R is the *connecting rod;* R S the crank; S the main shaft, on which is firmly fixed the *fly wheel* V V.

The above are the essential parts of the engine, each of which shall be described in detail as far as necessary. The other parts are the *governor*, to open and shut the *throttle valve* in the *steam pipe*, the *slide* and slide casing, the starting gear, the parallel motion, the eccentric, etc.

(1) **The Beam** is a lever of the first kind, and needs no description after an examination of the figure. The power is conveyed into the cylinder which moves the piston, the weight is the force conveyed by the crank, the fulcrum is the main centre.

(2) The piston, the cylinder, the air pump, condenser, and stuffing box have been already described.

(3) **The Feed Pump** is a force pump with a plunger to force the water into the boiler.

A is a solid plunger, v, v' and v'', are three valves; $b\,v''$ is the pipe that brings the water to the feed pump; $c\,o$ carries away the waste; $C\,c$ leads to the boiler, while c is a cock to shut off the feed from the boiler.

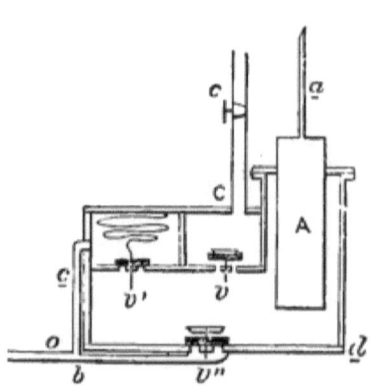

FEED PUMP.

It acts thus let us suppose the plunger is raised up, then a vacuum is left in the valve box $c\,d$, therefore water rises through the suction valve v''. Let us suppose $c\,d$ is filled, then the descent of the plunger will force the water through the delivery valve *v*, and up the pipe C *c* to the boiler. But suppose the cock *c* should be closed, then the great pressure of water will force back the strong spring and open the valve *v'*, so that the water can pass down *c o*. Sometimes, instead of this arrangement for the waste water, the pump rod is disconnected when no feed is wanted, and thus the power necessary to work the pump is saved.

(4) **The Pump** is an ordinary pump for raising water.

(5) **The Connecting Rod and Crank** have been already partially described. They are used for converting a rectilinear into a circular motion. The connecting rod should be as long as possible; it is generally from three and a half to four times the length of the stroke, but when cramped for room, or otherwise, a much shorter rod is made sufficient. The longer the connecting rod the greater its advantage. It has more leverage, and therefore does more work. A short connecting rod gives much pressure upon the guides and a great strain on the crank and crank pin, but with a long connecting rod this pressure and strain are avoided.

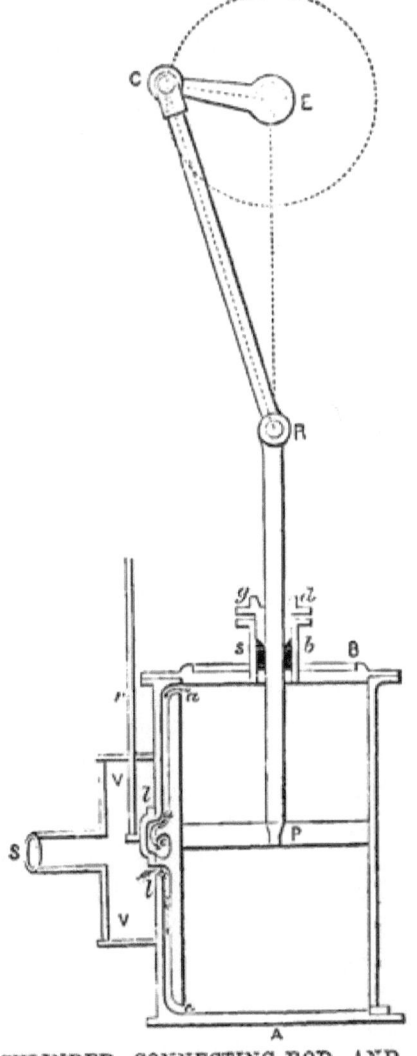

CYLINDER, CONNECTING ROD, AND CRANK.

With a short connecting rod it is difficult to properly adjust the cut off.

72. Cylinder and Crank.—The figure on previous page is a representation of a cylinder with a locomotive or three-ported slide. Cylinders are constructed of cast iron, and bored with the nicest precision. They must be perfect cylinders, the same diameter from end to end.

A B is the cylinder, P the piston, and P R the piston-rod; C E the crank, and E a section of the main shaft turned by the crank, and connecting rod C R; $s b$ is the stuffing box, and $g d$ the gland; $l l$ is the slide, and r the slide rod by which the engine moves the slide up and down; S is the end of the steam pipe which brings the steam from the boiler to the cylinder; a is the upper port, c the lower port, e is the exhaust port by which the steam escapes from the cylinder to the condenser after it has done its work.

73. How the Engine is Worked.—Suppose the slide is in the position shown in the figure, and that steam fills the valve chamber V V, through the steam pipe S. Now, it cannot pass the back of the slide into the upper port a, because the slide is covering it over; neither, for the same reason, can it pass to the exhaust e; but it can pass into the lower port c, in the direction of the arrows, and drive up the piston P, while, as the piston goes up, the steam that drove it down and filled the cylinder on the upper side above the piston is escaping freely through a, in the direction of the arrows, and passing off to the condenser through e the exhaust port.

When the piston has arrived at the upper end of the cylinder, or at the top of its stroke, the slide $l l$ has moved down lower, so that the lower port c is closed against the admission of steam, and the upper one a opened; therefore steam will enter the upper port and escape at the lower, in a contrary direction to the arrows, the piston returning to the bottom of the cylinder.

74. Slides.—The *locomotive slide* has been already partially described, when speaking of the beam engine and the way the steam is admitted to the cylinder. The various slides used are the long D, short D, Seaward's, cylindrical, gridiron, etc.

75. The Locomotive Slide is represented in the annexed figure, in which the dark shaded parts are the slide, and the ports are marked *port*. *c* leads to the condenser. The whole of the drawing is covered over by the slide casing, and steam is brought to the back of the slide at A by the steam pipe (not shown). When the steam is acting, it is clearly seen that it presses with great force against the back of the slide at A. The valve rod is shown attached to the back of the slide. When in the position as given in the figure, it is quite evident no steam can pass into the ports and go to the cylinder, as they are both covered over; but when the slide rod moves the valve up, the steam can pass into the lower port, and drive the piston up, while the steam that is in the upper part of the cylinder can come out at the upper port, when the form of the slide compels it to pass into *b b'* and through *c*, which leads to the exhaust, hence *c* is called the exhaust port. When the slide comes down again, both ports are first closed, then the upper one is open to steam and the lower one to the exhaust, precisely the reverse of the first case. As there are three ports, two steam ports and the exhaust port, this valve is sometimes called the "three ported slide."

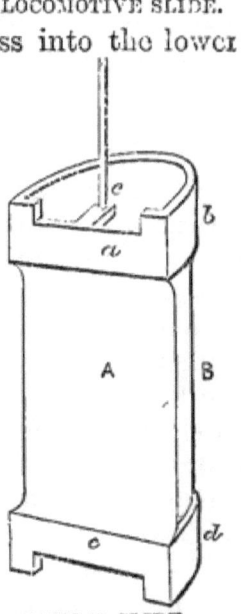

LOCOMOTIVE SLIDE.

LONG D SLIDE.

76. The Long D Slide is so called

because its cross section forms the letter D. The two faces, *a* and *c*, fit against the ports. The body, or waist, A B, is smaller than the parts *a b* and *c d*. The steam comes along the steam pipe, and can pass freely round the waist of the valve, and pressing against both back and front it is almost an equilibrium valve. The steam cannot pass by *b*, *d* nor *a*, *c*, because the two former parts fit closely to the slide casing, and the two latter press against the ports; only when the valve A is lifted or depresssd can the steam enter the cylinder from round the valve. When the steam comes out of the upper port it passes right down the slide at *e* to the exhaust. This is the peculiarity of the slide, that the exhaust passage from the upper port is through the valve.

77. Short D Slide may be described as consisting of the upper and lower portions *a* and *c* of the long D, but the passage is closed, and they are joined together by a rod. The steam is still brought to the waist, but cannot pass either *a b* or *c d*, unless the slide be lifted up. Its action is somewhat similar to that of the long D, excepting that the way to the exhaust is not through the slide. There are separate exhaust passages from the top and bottom ports.

78. Seaward's Slides were first used by the inventor, after whom they are named. There are four slides, two for the exhaust and two for steam. A is the steam side of the cylinder, and B the exhaust side. When the slides are in the position shown in the figure, the piston is ascending. Steam enters at C; the upper port *a* being closed it cannot enter the top of the cylinder, but it can enter at the lower port *b*, and drive the piston up. As the piston ascends *c* is closed and *d* open, so that the steam which drove the piston down is escaping through *d*. When the piston is descending, *a* and *c* are open, and *b* and *d* closed. D is the way to the exhaust, and B is called the exhaust side of the cylinder; *a* and *b* are termed the induction ports, *c* and *d* the eduction. The slides are kept against the face of the ports by springs, so that any water that enters the cylinder through priming can easily escape.

79. Cylindrical Slide.—These slides have been introduced and fitted to engines by Maudslay & Field. They are cylindrical in shape. The slide faces are hollowed out concave, and fit on convex nozzles. They are placed between the two cylinders, being used in double cylindered engines, and, when raised, the steam is admitted to the top of the cylinders, and the down stroke follows; and, when depressed, steam enters beneath the piston, and the up stroke is effected.

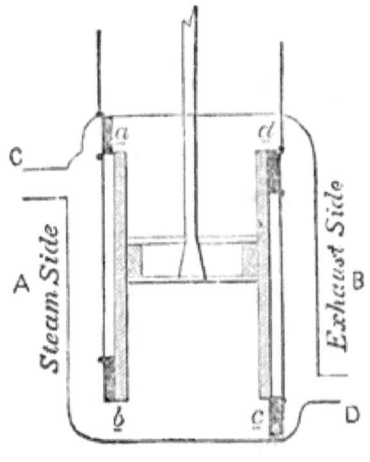

SEAWARD'S SLIDE.

80. The Gridiron Valve.—The gridiron valve is one of the most effective contrivances to give a large opening for steam by a very short movement. Each port is sub-divided into two or more narrow ports, while the valve face has openings to correspond. The principle is the same as that of an air grating in the floor, we have only to give the top plate a slight motion when it is open or shut; the same with this valve, except that the motion is rectilinear and not circular. If A B represent the ports of the cylinder, and the dotted lines the slide face, it is seen that, by simply lowering the slide (face) the smallest amount, the upper ports, A, are immediately open, and the lower, B, closed. When the

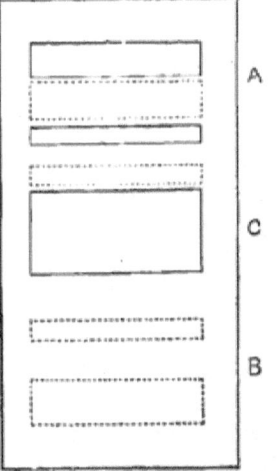

GRIDIRON VALVE.

slide is pushed back, the lower ports will be opened and the upper closed.

Full Steam is the position of the valve when fully open, and the piston is continuing its motion.

Cut-off is the position of the valve when it has just closed the port against the admission of steam.

Angular Advance is the angular measurement of the arc described by the centre of the eccentric while passing from the place it occupies when the valve is at half stroke, to that which it occupies at the commencement of the stroke of the piston.

Linear Advance is the distance which the valve moves while the centre of the eccentric is describing the above angle.

81. **Lap and Lead of the Locomotive Slide.**—The width of the opening of the steam ports, for the admission or for the release of the steam at the beginning of the stroke, is known as lead. On the steam side of a locomotive slide, it is known as outside lead, or lead for the admission; on the exhaust side it is inside lead, or lead for the exhaust. When the valve is placed at half stroke over the ports, the amount by which it overlaps each steam port, either internally or externally, is known as lap. On the steam side it is named outside lap; on the exhaust side, inside lap. When the terms lap and lead are employed, they are understood to refer to outside lap and lead only.

The advance of the eccentric is a term used to denote the angle which it forms with its position at half stroke, and when the piston is at the commencement of its stroke.

The locomotive slide, as seen in section in the following figure, has neither lap nor lead, but did it extend to the faint dotted lines $b\ b'$, it would have lap on the exhaust side to both ports; while, on the contrary, if it reached to the dotted lines $a\ a'$, it would have lap on the steam side. Lap is chiefly used on the steam side. To see what effect this will have, let us examine the top port, and suppose the slide going up. It is evident if the slide reaches to the dotted line *a*, as it rises from the bottom of the upper port, it will close it sooner against the admission of steam than it would be otherwise if the slide were constructed simply as drawn in the figure; therefore the steam that has had time to get into the cylinder has to perform the rest of the stroke expansively.

Lap on the exhaust or eduction side, $b\ b'$, is always less than that on the steam side, and closes the port to the exhaust sooner than it would otherwise be, and thus prevents all the steam from rushing out to the exhaust: the steam remains behind, and the piston acts against it as against a cushion, and all sudden jar and stoppage is avoided. Sometimes there is no lap, and even *less* than *none*, or negative lap; then the valve cannot cover both ports at once. When the slide has neither lap nor lead, the breadth of the slide face is equal to that of the steam port, and the travel of the slide twice the breadth of the port; but when the slide has lap, the travel of the slide must be double the lap with double the breadth of the steam port.

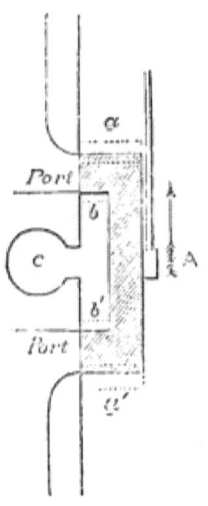

LOCOMOTIVE SLIDE.

82. **Lead.**—Let us suppose that, at the instant the *piston* is at the top of its stroke, the slide is in the position shown in the last figure, but that it extends only to the top darkly dotted line, then the port at that instant would be open for the admission of steam: this is what is called the lead of the slide. Remember the *lap* is when the *slide* is at its middle position, but *lead* when the *piston* is at the end of its stroke. The lap and lead of the D slide are explained in precisely the same way, but the steam slide is the inner and the exhaust the outer. There is always more lead required in

engines that are driven at great speed, than in those which work slowly. Again, in engines that travel fast, it is best to open the exhaust passage before the end of the stroke, or else the cushioning will act injuriously.

83. To Reverse the Engine with the Single Eccentric.—When an engine is fitted with a single eccentric, the engine is reversed by hand. The engineer notices whether the piston was moving up or down; if *moving up*, he takes the starting bar and admits steam to the *top* of the piston, so that it immediately descends, and the shaft begins to move in an opposite direction. The eccentric is fitted on to the shaft, so that it can be moved halfway round, or rather there are two stops on the eccentric, and one on the shaft. The shaft revolving, as we have just said, moves without the eccentric, so that the stop on the shaft leaves one of those on the eccentric, and when the shaft has moved halfway round, it comes against the second stop on the eccentric, which will be then in its proper position for working the slides, and so the motion of the engine is continued. To throw this eccentric in and out of gear, a recess is cut in the eccentric rod (care being taken that it is in its exact position), to this a pin is fitted to connect it with the slide rod or gab-lever pin. When the engineer has started the engine by hand (by lifting up the slide with the starting bar), and wishes to attach the motion of the eccentric to it, he watches his opportunity and lets the rod fall on the pin; the pin will in half a stroke fall into the recess. It is kept in its place by a bar or strip of iron placed over the entrance of the recess, held there by a spring.

84. The Double Eccentric, or Stephenson's Link Motion.—This contrivance, used both in the locomotive and marine engine, was invented by Stephenson to enable the engineer to quickly reverse his engine, and so go backwards or forwards at pleasure.

It consists of two eccentrics, H and G, with their rods A D and C E, the one called the forward, the other the backward eccentric. The two are connected by a link,

STEPHENSON'S LINK MOTION.

D E, with a slotway in it. In the slotway moves the block p, fastened to the end of the valve rod a.

The bell crank lever, D E D, is to move the link up or down. When the forward eccentric is moved, so as to work the valve rod, it moves the slide, and the ship or locomotive goes forward; but when the backward eccentric works the slide rod, the engine is reversed. The link motion is thus a simple and effective mode of reversing the engine expeditiously, and almost without trouble to the engineman.

When we consider that the forward eccentric rod, A D, sends the engine one way, and the backward rod, C E, sends it the other, we see that the travel of the slide has been reversed, as it were. Again, if the pin and link be

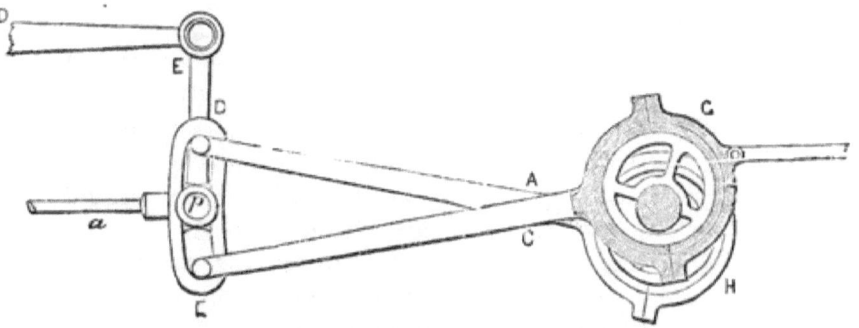

STEPHENSON'S LINK MOTION.

placed in the position shown in the figure, the slide has then but little travel, and we can see that this travel is increased just in the same proportion as the bell crank lever, D E p, moves the link D E up or down from the mid position. As the amount of opening for steam depends upon the motion of the slide, by leaving p in different positions in the slot we open and close the port at and during varying times. This is done by not placing the block at the extremity of the link, but at a distance from it, and resting the lever in its proper place. For this purpose an arc or sector with notches in it is attached to the link motion, to fix the handle in and secure the required opening the engineer may deem best for the speed required. This is not expansion, but rather wire-drawing the steam. In fact, Stephenson's link motion cannot properly be used to give different grades of expansion, it only alters the travel of the slide; for when the pin is in the middle of the link, the motion of neither eccentric is imparted to the slide rod. The pin being at the end of the link, the slide rod will receive full motion, and full steam will be given to the cylinder; but when the block lies nearer to the centre of the link, less and less steam is given to the engine, and consequently it moves the more slowly. This point is more completely illustrated under the heading "The manner in which the Link Motion distributes the Steam," in the chapter on the Locomotive Engine in the Advanced Work on *Steam* in this Series.

85. **Expansion Gear for Marine Engines.**—Various plans are adopted by different makers. Some use cams placed on the shaft in such a position that, when the valve is connected with the cam, by an arrangement of rods, levers, etc., steam can be admitted into the cylinder, but when not so, the ports are closed against the admission of steam. The great objection to this arrangement appears to be, that when the roller comes off the cam, it, together with the valve, drops with a sudden jar, which causes a very unpleasant noise in the engine-room, and also a great amount of wear and tear in the machinery itself.

The best plan appears to be to have an eccentric, to which is connected a sliding valve in the steam chest. This eccentric is fixed to the shaft in such a position, that when the valve is in connection with it, it shuts off steam at the required portion of the stroke. The different grades of expansion are regulated by a lever with recesses in it. This is among the connections of the expansion gear. Care is taken, when throwing it out of gear, that the expansion valve is not closed, or else the engine will stop. In some cases the throttle valve is used as an expansion valve, under which circumstance the full benefit of expansion is not gained, for that requires the total cut-off of steam, which the common throttle valve cannot do on account of its shape, but it wire-draws the steam.

86. A Method of Carrying out Expansion. — The principle of expansion is now carried out in all engines. The end is attained either by *cutting off* the steam from the cylinder by means of lap on the slide valve, or by valves called *expansion valves* preventing its ingress to the slide jacket. The expansion valve is so placed that the steam has no communication with the slide jacket except through the valve.

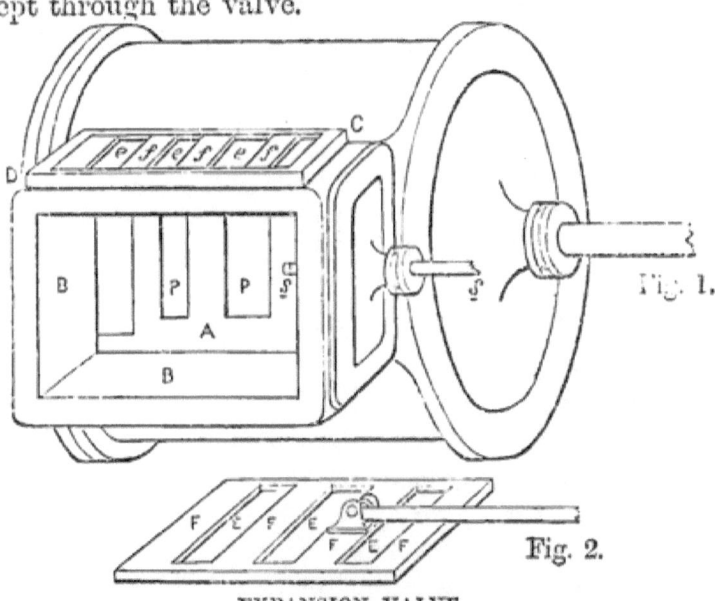

EXPANSION VALVE.

The valve now in general use in marine engines, and which is finding most favour with engine-makers, is the gridiron expansion valve; its construction and name being easily understood from figure 2. It consists of a series of ribs F F F, between which are the apertures E E E, through which the steam passes when these apertures are over those of figure 1 marked $e\,e\,e$. For the gridiron valve in figure 2 works on the face of D C, the ribs F corresponding to f, and opening E to e. The steam passes through E e, E e into the slide jacket B B below, and thence through the ports P P, etc., into the cylinder, when the slide valve, of which ss is the rod, is in the proper position to allow it to do so. If this slide valve covers the ports P P, of course the steam cannot pass on; ss with its valve receives motion from the ordinary double eccentric, but 2 is moved by the expansion eccentric figure 3. The expansion valve is of this particular construction, the gridiron, to give a large area to the opening with a short stroke of the valve; for, with a short stroke equal to the breadth of one of the ribs or faces F, we have an opening $= E + E + E$, or $e + e + e$. The motion of the expansion valve (2) is derived

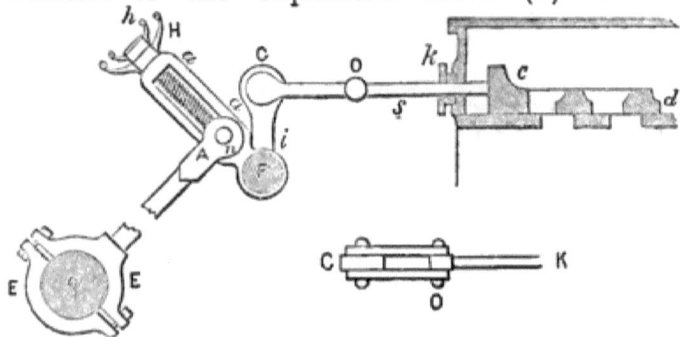

Fig. 3. WORKING OF EXPANSION VALVE.

from an eccentric keyed to the main shaft, and is transmitted to the valve by means of the simple arrangement in figure 3. We will state again how the steam passes to the cylinder. The top or cover of slide casing is wholly removed in our figure 1, the end of the steam pipe comes to C D, but the steam cannot pass to B B until the valve (fig. 2), being moved, will allow the steam to pass into B B, even then the steam will not enter the cylinder if the ports are covered by the valve (which is removed from the ports A A), and of which the spindle is ss; but when this valve is shifted so as to uncover the ports, the steam has free ingress to the cylinder. By this plan it is evident that valve 2 will give any amount of expansion by stopping the supply of steam to valve A (removed).

Fig. 3. Here E E is an eccentric keyed on the shaft S to work the levers of the valve. H is a handle to screw up or down the block through which the pin n passes by means of a square threaded screw, as seen in the figure, working in the block at n, and h is to jam or keep it in its place. When n is moved nearer the handle, the grade of expansion is greater; when screwed down, the opposite is the case. F is the weigh shaft supported by two bearings, not shown in the figure. G i is a short lever answering to the gab lever keyed on the shaft F; to the end G is attached a pair of links to allow for the motion of G not being in a straight line, of which G O is one; to the end O of these links is fastened the valve rod O k. The end of the eccentric rod n moves H F backwards and forwards on the centre F, which moves G i on the same centre, and this gives the reciprocating motion to G O k, and the slide with the requisite amount of expansion. The grade of expansion is marked along the edge at aa. While the arrangement here described is giving the required amount of expansion, the ordinary link motion is admitting the steam, or cutting it off from the cylinder (fig. 1) by means of the valve removed from A.

87. The Amount of Water Required for Condensation.—The proper temperature at which to keep the condenser is as near as possible 100° F. or 38° C. At this temperature the steam is sufficiently condensed, while the air pump has relatively the least quantity of water to raise; or, with a maximum amount of useful condensation, we have a minimum amount of water to lift.

Let us suppose the condenser is to be kept at 100° F., and the temperature of the condensing water is 50° F., then out of every unit of water 100° − 50° = 50° of cold are available to condense the steam.

Watt assumed the total heat in steam to be 1112° F. (latent and sensible heat of steam we have called 637°·2 C. or 1147° F.); therefore there are 1112 units of heat to be overcome, which will take $\frac{1112}{50} = 22.24$ units of water; or it will take $22\frac{1}{4}$ more times water than is turned into steam. As a cubic inch of water produces a cubic foot of steam, it will take $22\frac{1}{4}$ cubic inches of water to condense one cubic foot of steam.

Watt allowed 28·9 cubic inches, or about a wine pint, for every cubic inch evaporated.

In this calculation we have given the result arrived at by Watt. We will now perform the calculation, using degrees centigrade, making allowance for *the heat which will be left* in the condensed steam, and using the more accurate number, 637°·2 C.

Suppose the temperature of the condenser is to be maintained at 38° C., and the temperature of the condensing water is 10° C., what amount of water will be required for condensation?

The total amount of heat in a given unit of steam is 637·2 units C.

The amount imparted to each unit of water is $38 - 10 = 28$ units C.

Of the 637·2 units of heat in each unit of steam, it must give up $637·2 - 38 = 599·2$ units.

∴ the units of water required $= \frac{599·2}{28} = 21·4$.

Or, a cubic foot of steam, as it is produced (very nearly) by a cubic inch of water, will require 21·4 cubic inches of water to condense it. More is always allowed, because it is impossible so to arrange the condenser that every drop of water shall at once consume its allotted amount of heat.

The temperature of the condenser will always give an idea as to the vacuum. If the temperature of the condenser is above 100° F., then more water must be supplied for condensation; if it is below 100° F., then the cocks must be closed a little, as too much water is being used, and the air pumps will have too much work thrown upon them. When the air pumps are labouring too hard, it is one sign that too much condensing water is being used. A thermometer therefore inserted in the condenser will show the state of the vacuum. Generally, the engineman trusts to his vacuum gauge to tell him the state of his condenser. If the vacuum gauge is low, too little water is being used, and he must remedy the defect accordingly.

88. Blow-through Valve.—The blow-through valve of an engine is used to drive out all water from the cylinders, casings, and condensers before starting. It is placed at the bottom of the slide casing so as directly to communicate with the condenser. But sometimes, in the case of the locomotive, one is placed at each end of the cylinder, and worked by a handle from the starting platform. Some engine-makers fit a small locomotive slide and ports for the purpose, which can also be used to start the engines. Before the engine begins work, steam is admitted through the blow-through valve, and the cylinder first cleared of air and water; the steam passing on clears the condenser in the same way, so that, as soon as the engine starts, a good vacuum is obtained in the condenser. This last is the chief object for which blow-through valves are fitted.

S P is the steam pipe; the steam having been brought to the back of the slide cannot enter the cylinder unless the long D slide be lifted up or down, neither can it go to the condenser unless the blow-through valve B be opened by means of the handle h. When the valve B is lifted off its seat, then steam can freely pass to the condenser, and blow out all air and water that may be in it; when no blow-through valve is fitted, by the tedious process of alternately letting the steam pass to the top and bottom of the cylinder, and by raising and lowering the slide, the steam may be sent to the condenser, from which it will in time expel the air and water.

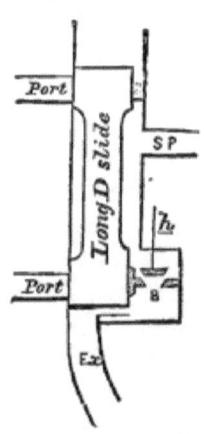

BLOW-THROUGH VALVE.

89. Mercurial Gauges.—Mercurial gauges are and have been used to show the pressure of steam and the vacuum. But as they are very cumbersome, and nearly obsolete, it is useless to describe them, but we may say this much—

(1) **The Long Barometer Gauge.**—The pressure of air corresponds to very nearly 30 inches of mercury, which being about 15 lbs., 2 inches of mercury indicate 1 lb. pressure. A bent tube in the shape of a U, partly filled with mercury, was taken, and one end inserted in the boiler; as the pressure of steam increased, it would drive the mercury down the one part of the tube, in communication with the boiler, and up the other; a graduated scale of 2 inches to the pound showed the pressure of steam in the boiler.

(2) When used as a vacuum gauge, the mercury would follow the vacuum, and rise up the part of the tube connected with the condenser.

(3) **The Short Barometer Gauge** was used to show the vacuum. It was of similar construction to the last; but between the legs, communicating with both, was a reservoir of mercury. As the pressure was taken off the reservoir the mercury fell down one arm, which was short; for as the vacuum between 10 and 15 lbs. only was wanted, the arm was made short, and would remain full of mercury till the pressure fell to 5 lbs. only; so that when the mercury stood 10 inches high, we should have 5 lbs. pressure of air in the condenser; when 8 inches high, 4 lbs., etc.

The mercurial or barometer gauges are old-fashioned, and are hardly used now or fitted to new engines; therefore we have given no figures, merely a short description of them. To these gauges there are scales graduated to every two inches, so that by looking at them the engine-man can tell at a glance the condition of his vacuum. If the mercury stands at 20 inches, then there is $\frac{20}{2} = 10$ lbs. vacuum, or $(15 - 10 =)$ 5 lbs. pressure of air in the condenser. If the mercury stands at 24 inches, there is a vacuum of $\frac{24}{2} = 12$ lbs., or the pressure of air in the condenser is $(15 - 12 =)$ 3 lbs. Another form of vacuum gauge is this: An iron tube is fixed into the condenser and bent upwards. At the bottom near the condenser is a cock, to open or close the communication with the condenser.

Just above the cock is a small bowl for holding mercury, the tube passing right through the bowl, so that the mercury is round the bottom of the tube and outside it; the top of the tube is open. A glass tube, open at the bottom and closed at the top, a little larger in the internal diameter than in the outside diameter of the iron tube, is taken and placed right over the iron tube, the open end coming down into the mercury. When the communication with the condenser is opened, there being a vacuum within the iron tube, the pressure of the air on the outside pressing on the mercury will cause it to ascend between the two tubes; and, of course, the higher it rises the better the vacuum. It will ascend two inches for every pound. It is graduated, and a scale placed by its side; but as the mercury will sink in the bowl as it rises between the tubes, a pointer or piece of wire is attached to the scale, the end of which, bringing the scale lower with it, must be placed on a level with the mercury before the state of the vacuum is read off. Unless this precaution is taken, the reading is liable to error.

90. **Fly Wheel.**—The fly wheel is an accumulator of power, and assists the crank over the "dead centres." When the crank and connecting rod are in one straight line, as they must be twice in each revolution, the crank is said to be on its dead centre, because there the force of the piston is dead or ineffective. It is evident that when the crank is at right angles to the connecting rod, that the latter has most power on the former, but when the top or bottom dead centre is reached there is no reason why it should not remain there; but the action of the fly wheel then shows itself, for having on it a certain accumulated velocity, it cannot stop, but goes forward, carrying with it the crank over the dead centre. We thus have through the momentum of the fly wheel no perceptible variation in the velocity of the engine, but the unequal leverage of the connecting rod is corrected, producing a steady and uniform motion. The fly wheel, it must be remembered, is a *regulator* and reservoir, and not a creator of

motion, and when no fly wheels are used, as in marine engines, we must recollect that smoothness of motion is not an absolute requisite, and that the momentum of the engines themselves carries the cranks over the dead centres; but far more generally a pair of engines work side by side, whose cranks are at different angles, so that one assists the other at the critical moment. The accumulated velocity of the fly wheel, where the motion is required to be excessively equable, should be six times that of the engine when the crank is horizontal. The efficiency of the fly wheel, in producing uniformity of velocity, is materially modified by the motion of the machinery which the engine is required to drive, and regularity of motion is of much greater importance in some cases than in others; so that in proportioning a fly wheel to a given engine, attention must be paid to many particular circumstances which cannot be given in a general rule.

91. Equilibrium Valves. — Equilibrium valves are those upon which the steam presses with equal force (or very nearly equal force) both upon the top and bottom, being ready to move easily when required. The following figure will give a good idea of an equilibrium valve:—

S is the steam-pipe, through which steam is introduced into the valve-box A B; a and b are two conical valves on one valve spindle $c\ d$, kept in its place by the socket d. The steam is required to pass at intervals along C. This it will do with full force when the valves are but slightly lifted upwards. It is seen that if a and b be very nearly equal, the valve **is in**

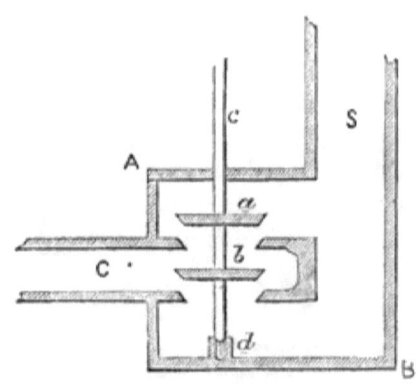

EQUILIBRIUM VALVE.

equilibrium, and only a small force is required to lift it, for the pressure of steam on the top of a is counteracted by that on the bottom of b.

92. Cornish Equilibrium, Double-beat, Crown or Drop Valve.—A B is the valve-box. Steam enters it, let us say, from C, and is required to go along D, after passing the valve. It might with equal propriety be supposed to come from D and be passing down C. The part drawn with cross lines or section, is a cylindrical piece of iron fitting down on two rings, $b\ b$ and $b'\ b'$. The small squares are the sections of the rings; suppose these to go all round. It is evident that when the valve is down on the rings no steam can pass, but as soon as lifted it can rapidly pass through the two openings marked a in the paths indicated by the arrows. These openings extend all round in a circle. A very slight movement gives a large opening for steam. The seats $b\ b$ and $b'\ b'$ are called the beats. Sometimes these valves are made with three or four beats.

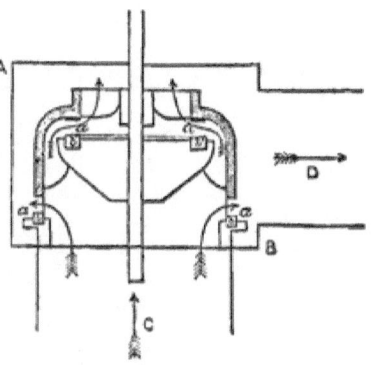

CORNISH DOUBLE BEAT VALVE.

93. Escape Valve.—The escape valve should have been noticed when describing the cylinder. They are fitted in the top and bottom of the cylinder, being kept in their places by weights or springs. Water that gets into the cylinder through condensation or priming, as it is incompressible, would inevitably break something, were not provision made to allow it to escape through the escape valves. They are loaded with a weight or spring greater than the pressure of steam in the boiler. *Test* or *pet cocks* are also fitted to the tops and bottoms of the cylinders in marine engines for the same purpose. They are opened on starting the engine, and shut when properly under way. The escape valves are *always* ready to act, and are held in their places by weights, which keep them closed only so long as the pressure in the condenser is below that in the boiler.

EXERCISES CHIEFLY FROM EXAMINATION PAPERS.

1. In what way is steam admitted into the cylinder. How is the apparatus worked (1865)?

2. How is steam admitted into the cylinder? Describe with a sketch the usual mode in marine engines for working the gear connected with the slide (1868).

3. Describe with a sketch some form of slide valve as connected with the steam cylinder of engine, and explain its action (1869).

4. Draw in section the cylinder and slide valve of a double acting engine, and explain the manner in which the valve regulates the admission and exit of the steam.

5. Describe with a sketch the single acting engine (1871).

6. Describe the long D slide (1867).

7. What is the use of the expansion valve? Show by a diagram the pressure of the steam in different parts of the stroke when worked expansively (1867).

8. Describe the barometer gauge in common use (1867).

9. What is meant by the terms cushioning and clearance?

10. Does the amount of clearance above the piston of a side lever engine usually increase or diminish as the engine wears (1868). See next chapter.

11. Describe the Cornish double-beat valve (1868).

12. Describe the method of working a slide valve by an eccentric (1869).

13. There are three valves connected directly with the steam cylinder in Watt's single acting condensing engine, name them. During what portions of the up and down strokes of the piston should these valves be respectively open or shut? and for what reason (1871)?

14. Valves used to close a passage through which steam or water under pressure may be required to pass are so constructed as to be capable of being lifted against this pressure with a very small expenditure of force—sketch a valve of this kind, and explain its action (1871).

15. Describe some form of slide valve as fitted to the steam cylinder of a double acting engine. Sketch the valve in section

with the opening over which it slides, and give it some amount of lap on the steam side. How is the face of such a valve made truly plane (1871)?

16. For what purpose are escape valves fitted to the cylinders of marine engines? How are such valves kept closed, and what determines the least amount of load which must be put upon them (1871)?

CHAPTER VI.

THE LOCOMOTIVE ENGINE.

History of Locomotive—Stephenson's Engine: The "Rocket"—General Description of a Locomotive—Crampton's Engines—Tank Locomotive — Bogie — Locomotive Boiler — Shell of Boiler—Through Tie Rods—Tubes—Clearance—Fire Box—Staying the Furnace—Fire Bars—Ash Pan—Smoke Box—Blast Pipe—Heating Surface—Safety Valves—Chimney—Damper—Steam Dome—Man Hole—Regulator—Whistle—Pressure Gauges—Salter's Spring Balance—Reverse Valve—Bourdon's Gauge—Sector.

94. THE locomotive engine is the product of many minds. The early workers at it were Trevithick, Hedley, Murray, Hackworth, and Stephenson; to the latter is usually assigned the greater part of the credit given for its introduction. He fortunately won a £500 prize offered by the directors of the Liverpool and Manchester Railway, for the best locomotive. This quickened the energies of an active and powerful mind, and the result is that, with the inventive genius of others to assist in the work, we have to-day the wonderful pieces of mechanism, called locomotives, tearing away all over our island at a speed of 50 and 60 miles an hour.

95. Stephenson's Engine: "The Rocket."—Stephenson found the locomotive a small clumsy engine, and after many trials and much experience left it almost the perfect machine we see it to-day. His first engine, made to "lead" coals from the pit, was constructed at Killingworth in 1814; it was supported on four wheels three feet in diameter; it had a wrought-iron boiler with a single flue, the fireplace was within the boiler, and the two vertical cylinders were half immersed in the same. The motion was conveyed to the wheels by a method that had previously been adopted by Hedley, who used cranks and toothed gearing. The cranks worked at right angles to each other, and the pistons made two strokes for each revolution of the driving wheel. As seen in the figure, the axle of each pair of driving wheels had a 24-inch toothed wheel keyed on to it, and the axles being 5 feet from centre to centre, they were geared together by three intermediate wheels of one foot in diameter. The centre

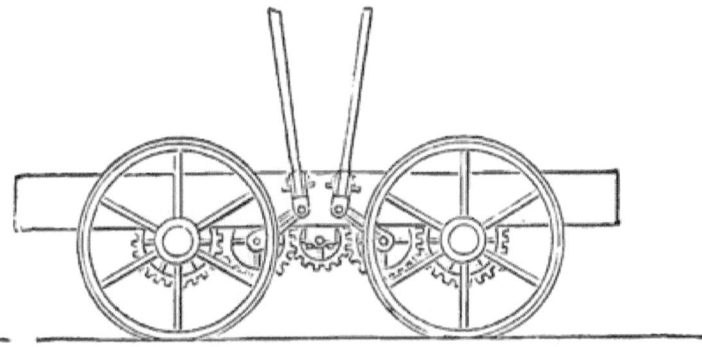

STEPHENSON'S DRIVING GEAR, 1814.

wheel acted as a regulator, and preserved the two cranks at right angles, and thus kept the propelling power in equilibrium. This engine did not answer very well, its radical defects were the single flue and the wide chimney; the waste steam does not appear to have been sent into the chimney. Stephenson soon abandoned the toothed gearing to convey the motion to the driving wheels, and introduced springs to carry the weight of the engine. Springs were first used by Nicholas Wood.

The annexed figure represents the "Rocket" as it appeared when it ran in the famous Rainhill competition.

It was a four-wheeled engine supported on springs, and with a supply of water in the boiler weighed 4 tons 5 cwt., with its tender loaded it weighed 7 tons 9 cwt. Its boiler, of which the accompanying figure is a section, was cylindrical, 6 feet long, with a diameter of 3 feet

4 inches; through it passed twenty-five copper tubes 3 inches in diameter; these conveyed the heated air, gases, and other products of combustion from the "fire box" at one end of the boiler into the tall chimney,

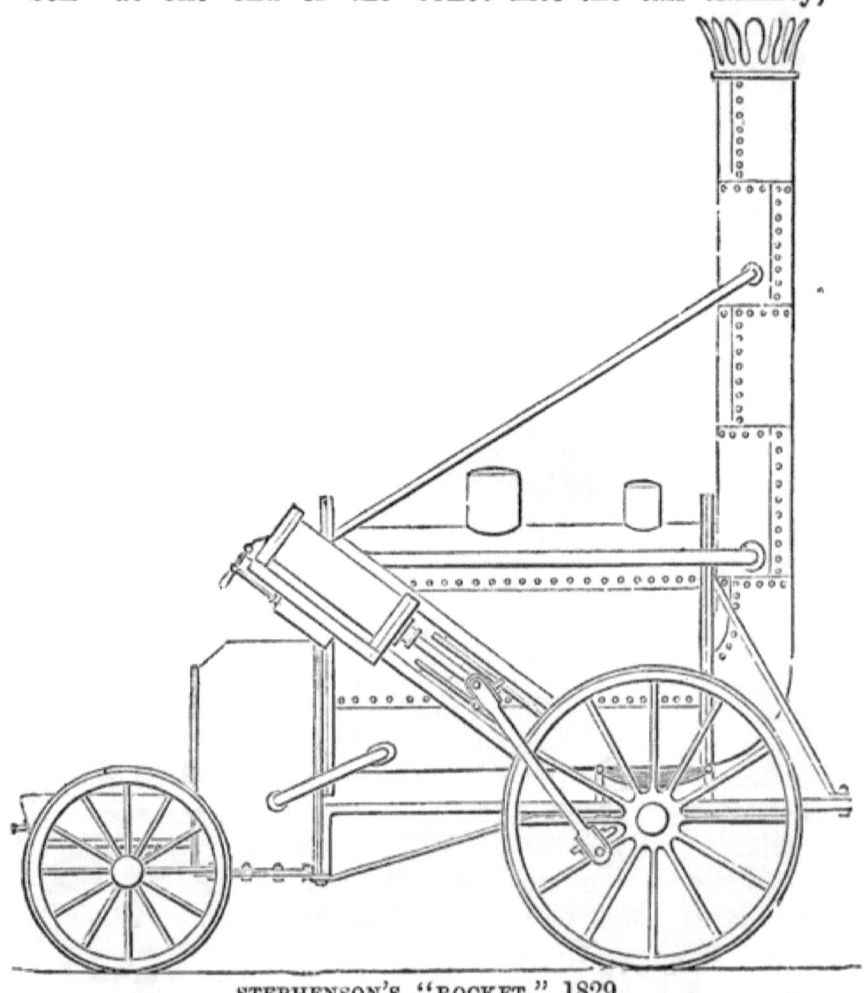

STEPHENSON'S "ROCKET," 1829.

12 inches in diameter, to the farther end, after passing from end to end of the flue. The heating surface of this multitubular boiler was 117¾ square feet; the use of these tubes gained Stephenson his victory, and laid the foundation of his fame. The body of the figure on

previous page is the boiler barrel with tubes inside. The fire box or furnace is represented on the left hand side close to the smaller wheel. It will be noticed that a small tube goes from the boiler barrel to the furnace, this was to allow water to run round the fire box casing; at the top of the fire box was another tube running into the boiler (in our figure it is omitted and hidden by the upper end of the cylinder), to allow the steam generated in the fire box casing to enter the boiler.

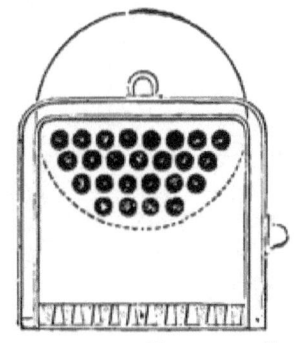

BOILER OF "ROCKET," 1829.

The safety valve is the projection on the top of the boiler nearest the chimney. The cylinders were two, one on each side; one is seen to the left just above the fire box, inclining to the rails at an angle of 45°; this was a poor arrangement, as the pistons slightly lifted the boiler up and down on the springs. It is seen that the connecting rods worked on crank pins on one of the spokes of the driving wheels, and thus the motion of an ordinary connecting rod and crank was gained. The diameter of the cylinder was eight inches, and the stroke $16\frac{1}{2}$ inches. The exhaust steam from each cylinder was carried through a pipe, and turned upwards into the chimney, but the exhaust orifice was not contracted.

96. Contrast between the "Rocket" and Recent Locomotives.—The cost of the "Rocket" was not to exceed £550; modern engines cost upwards of £2000. It weighed 7 tons 9 cwt. with its tender; the working weight of some modern engines and tenders exceeds 45 tons. The driving wheel was 4 feet $8\frac{1}{2}$ inches in diameter, cylinders 8 inches, and stroke $16\frac{1}{2}$ inches. Engines are now running with a driving wheel 9 feet in diameter, and cylinders 18 inches, and stroke 24 inches. The greatest speed attained by the "Rocket" on its trial was 24 miles an hour, for a distance of one mile and a half. Some of the express engines on the London and

North-Western Railway have attained a speed of 73 miles per hour between Holyhead and London. The pressure on the boiler was not to exceed 50 lbs. on the square inch when working, although the company were to be at liberty to test the boiler, etc., up to a pressure of 150 lbs. on the square inch. Now new locomotive boilers work at a pressure rarely less than 120 lbs. on the square inch, and in many cases 140 and 150 lbs.

97. General Description of a Locomotive. — Locomotives run generally on six wheels, though occasionally

more are employed. The large wheels are called the driving wheels, because they are driven directly by the cranked axle. The two wheels in front, or the front pair directly under the chimney, are called the leading wheels, and the pair near the fireman, the trailing wheels. The chimney is seen on the right hand, the furnace on the left, while the barrel of the boiler with the tubes lies between.

In this sectional elevation (Plate I.) F is the furnace, with f the furnace door; the furnace is seen surrounded by the outer fire box, but the screwed stays are omitted. Above and below B are the tubes running from the inner fire box to the smoke box S, two only are shown; around the

PLATE I

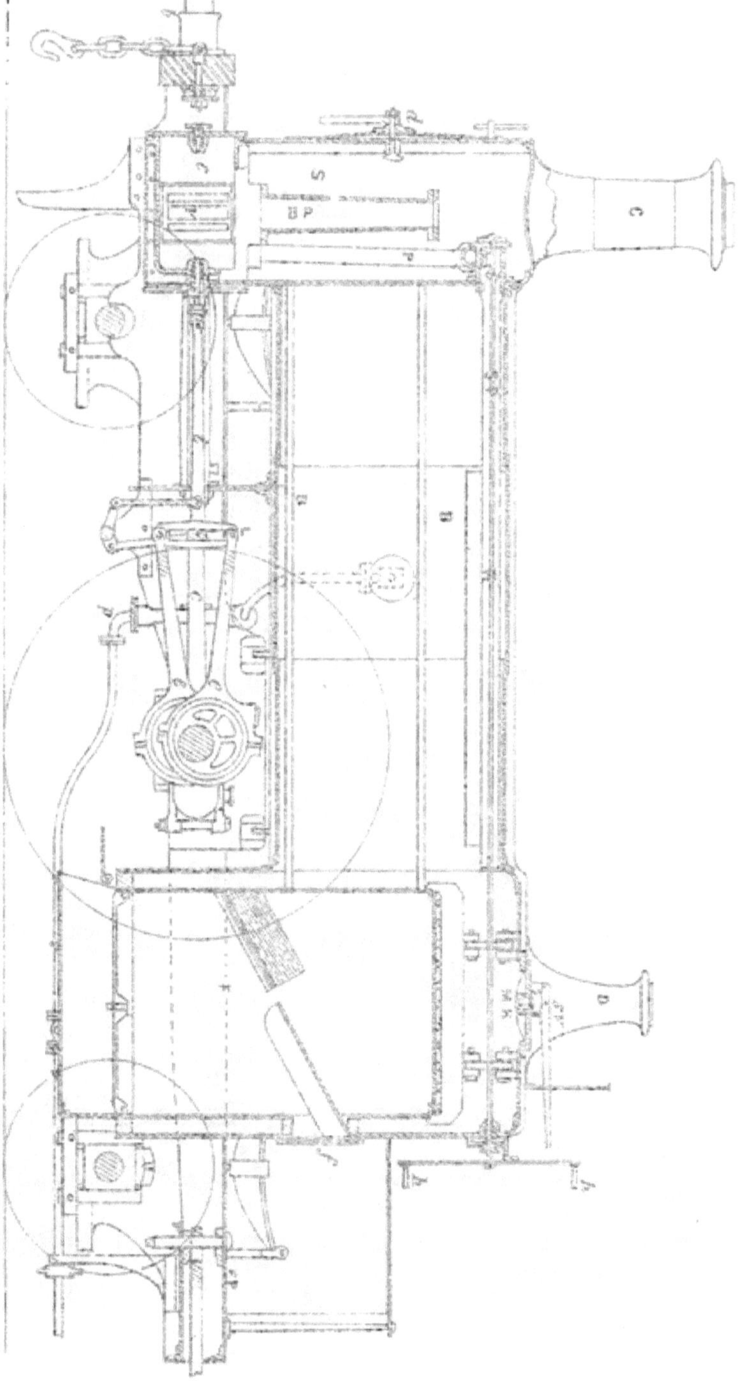

SECTION OF LOCOMOTIVE ENGINE

tubes and above them is the water ; the level of the water is called the water line. Admission to the smoke box is gained by a door at d ; this door is fitted as closely as possible to exclude all cold air. At the top of the smoke box S B is seen the chimney C, and within the smoke box is the waste steam pipe or blast pipe, B P, the mouth of which can generally be closed, or at least partially closed, to regulate the blast. The dome is at D, the steam from the boiler passes up D to the mouth of a pipe in it, this is the mouth of the steam pipe S P, generally closed by the regulator, which admits the steam to the cylinder; the regulator being opened, the steam passes along S P down the smoke box by way of P to the cylinder C, and sets the piston reciprocating; thus the engine is worked. In our figure the handle of the regulator is at h, and the regulator is not shown, the handle of course being worked by the engineman, who stands on the foot-plate, F P, at the back of the furnace; the whistle is also close to his hands, whilst one of the safety valves, S V, is under his control, the other he cannot interfere with. The man hole and man hole door are seen at M H, below the dome; the man hole door is taken off when it is wished to enter the boiler for examination, or to tighten the stays, etc. The large wheel in the middle is the driving wheel, turned by the crank, which is moved round by the connecting rod c, which is attached to the piston-rod i, the latter in its turn is firmly fixed to the piston. The front wheel next the chimney is called the leading wheel, securely fixed on the leading axle, and the wheel to the right the trailing wheel.

The following figure is another plan of arranging the locomotive. As said before, they generally run on six wheels, with the large driving wheel in the middle; but in Crampton's arrangement the large driving wheel is behind. In his engine circular motion is first given, by inside cylinders, to a cranked shaft, supported on bearings fixed upon the frame in the usual manner, and motion is communicated from this shaft to the driving wheels

96 STEAM.

behind the fire box by side rods. When outside cylinders are used they are placed midway in the length of the boiler, and connected directly to the driving wheel. The upper figure is Crampton's arrangement for outside cylinder, the lower for inside cylinders.

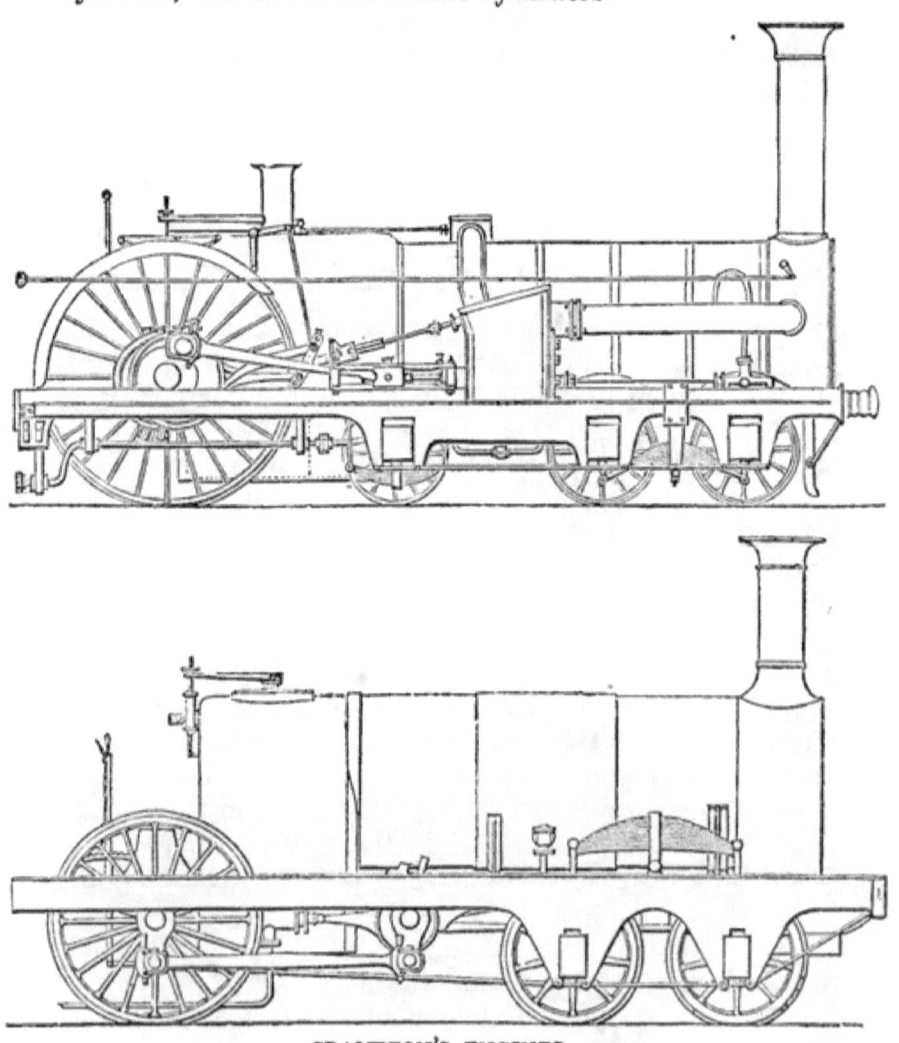

CRAMPTON'S ENGINES.

98. Tank Locomotives.—*Tank locomotives* are advocated in opposition to those of excessive weight to save the enormous dead weight, and are generally very light. They are constructed with a tank usually over the boiler, and occasionally at the sides, so that they can carry their own water, without being compelled to drag a tender after them, being independent of that seemingly fixed appendage.

99. Bogies.—The bogie is a truck on four wheels that will swivel round. Bogie carriages generally run on eight wheels. They were invented to meet the necessities of the American traffic, where, in passing through streets, it was sometimes necessary to turn round very sharp angles. Mr. Stephenson constructed the first bogie for America. "The engine was made two-wheeled, and a small truck on four low wheels supported the front end, being swivelled to it by a centre pin, or what the high road people call a perch bolt. This kind of truck, known in many places as a lorry, a trolly, and many other names, was, it appears, called in Newcastle a bogie, and the engine was therefore shipped as a bogie engine. It became the pattern or type for American locomotives."[*] When the engine or carriage is long, two bogies are employed with four wheels each.

100. Locomotive Boiler.—All locomotive boilers are of the class termed multitubular. They consist essentially of the barrel filled with tubes, while the two ends are named respectively the furnace, or fire box, and the smoke box. Boiler plates should be rolled from the best iron to about three-eighths or half an inch in thickness; these form the barrel, which has a diameter varying from three feet to four feet three inches in different boilers, and consists of three or six plates for each boiler, and their joints are so arranged as to give as much strength as possible.

The shell of the boiler is usually made of best Yorkshire, Staffordshire, or Lowmoor iron. The thickness of the plates varies from $\frac{3}{8}$ to $\frac{1}{2}$ of an inch, according to the diameter of the barrel of the boiler, which rarely exceeds 4 feet 3 inches inside. The joints are either lap or jump joints; if the first mode is adopted, they are made to lap

[*] Clark's *Railway Machinery.*

2 inches or $2\frac{1}{4}$ inches for single riveting; when jump joints are employed, 4 or $4\frac{1}{4}$ inch welts are applied to the seams, and secured to the boiler plates by two rows of rivets: the plates are or ought to be planed at the edges. The riveting is usually single, but for strength it should be in double rows in a zig-zag course. The rivets in size are from $\frac{3}{4}$ inch to $\frac{7}{8}$ inch in diameter, being placed at a pitch (from centre to centre) of from $1\frac{3}{4}$ inches to $1\frac{7}{8}$ inches. The barrel of the boiler is usually joined to the fire box and smoke box tube plate by a three inch angle iron. In the fire box shell, the front and back plates are joined to the others either by three inch angle iron, or by flanges turned on them to a four or five inch radius; the former is the simpler process, but the latter the stronger, fixing them more securely, and is the plan generally followed.

101. Through Tie Rods run from the smoke box tube plate to back of fire box; they are about one inch in diameter and four inches from centre to centre. Their number depends upon the size of the boiler. They are put in to stay the boiler, and assist the tubes in preventing the two ends from being blown out by the force of the steam.

102. Tubes.—The tubes may be of brass or iron; copper is too soft, brass is also better than iron for several reasons. It conducts the heat better, or communicates the motion of the fire more readily to the water than iron, and also resists the abrading action of the small coke carried through the tubes by the draught; it resists the action of impure water outside better, springs more easily under extra expansion, and is not so liable to break as iron is. Economically, brass tubes are at least as cheap as iron, as they will fetch, when worn out, half their original price for old metal. Tubes are fixed in the tube plates by widening with a mandril to fill the holes completely, turning over their protruding ends upon the plates. At the fire box end, ferules of wrought-iron, and in some cases of cast-iron, about an inch in length, slightly tapered, are inserted, and should, when driven, be left with about a ¼ inch projection into the fire box, so that should any of the tubes spring a leak on the road, they may be tightened by a tap or two from the end of a pinch bar. Ferules at the smoke box end are frequently omitted, which gives a free passage for small coal and cinders into the smoke box. Tubes are either of equal thickness throughout, or of a tapering thickness, from No. 9 wire-gauge at the fire box to No. 14 at the smoke box. Tubes wear unequally on the inside, and mostly at the fire box end. The first foot or eighteen inches should therefore be a little thicker than the rest of the tube. The number of tubes in a locomotive boiler varies from about 130 to 220. The distance between the tubes, called the *clearance*, is from ⅝ths to ⅞ths of an inch; but the larger the tubes the greater the clearance. The size of the tubes varies from 1¾ to 2 inches in diameter; they must not be too small, for fear of being choked, nor too large, for then the heating surface is diminished. If too small they are perhaps too numerous and crowded, when the water spaces are not of sufficient size to prevent priming, which is a serious evil if not effectually prevented in time; neither must they be too long, as the evaporative power of the heated gases rapidly diminishes as they recede from the fire box.

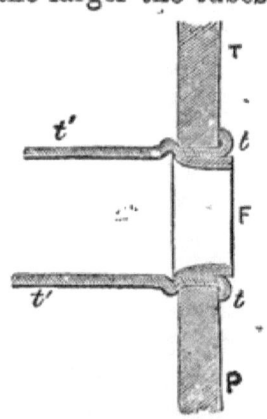

END OF TUBE AS SEEN IN FIRE BOX.

103. The Manner in which the Tubes are Fastened into the Tube Plates.—This has just been explained, and we illustrate it here:—T P represents a piece of the tube plate; *t t t t t'* is the brass tube, which, when driven in, projected a little beyond the tube plate, then the end was turned over on the plate as we see it at *t* and *t*; thus they are all left at the smoke box end, but at the fire box end they are further secured in their places by the ferules F.

104. Clearance.—*Clearance* is the space between the

tubes, and between the tubes and the boiler shell. It is required to allow a proper circulation of the water and steam around and between the tubes, and to give the steam plenty of room to rise, instead of remaining in contact with the tubes.

105. Fire Box or Furnace.—The *fire box* consists of two distinct parts, the external fire box, always made of wrought-iron, and the internal fire box, or furnace proper, of copper. The staying of the fire box is a question of the greatest importance, especially of that part immediately above the fire. Occasionally, the internal rectangular fire box is of iron, but copper is found to answer better, because it resists the intense combustion and conducts the heat more rapidly, and is not so liable to be burned away and ruptured at the thick lap joints and places where the sediment collects. The internal fire box is fastened to the external by screwed stays, screwed through both plates, and their heads left and riveted over. The space between the two is a *water* space.

In this figure (Plate II.) the part marked *b b*, etc., is the space between the internal and external fire box, the latter is seen in section on the outside, the former is seen inside the other; the short bolts running across are the screwed stays, many of the ends of which are seen at the front [*] of the fire box at *g g*, etc. The tubes B are marked by double circles above, there are about 178 of them in this boiler. The water spaces between the two fire boxes completely surround the inner fire box; it will be seen closed at the bottom by a square bar *c c*, which is bent and welded to the proper form to extend round the bottom of the inside fire box, and is riveted and tightly caulked to both fire boxes. The water in the water spaces is in free communication with the rest of the water in the boiler.

[*] The front of the fire box is what would be generally termed the back, *i.e.*, the *front* is the part *nearest the tubes*, so that the other side, where the *door* is, is the *back*. The engineman stands at the back of the fire box.

PLATE II

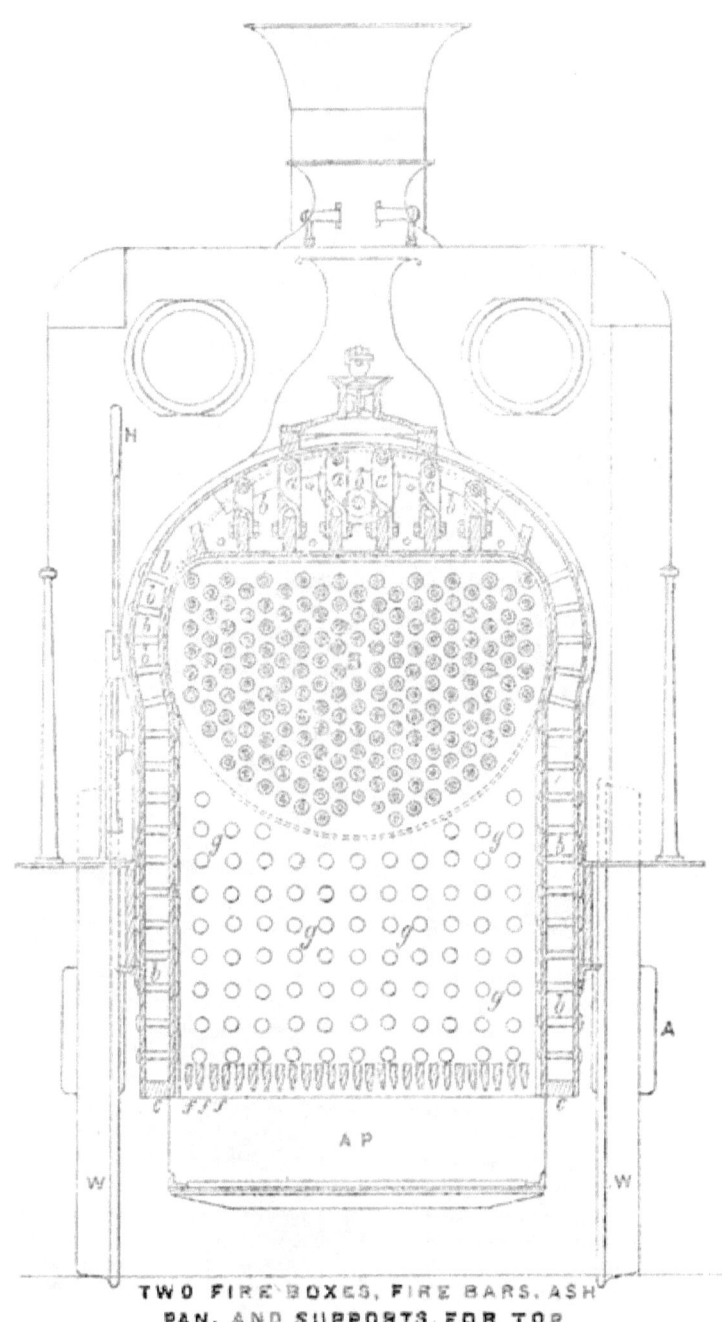

TWO FIRE BOXES, FIRE BARS, ASH PAN, AND SUPPORTS FOR TOP OF FIRE BOX.

FIRE BARS. 101

The *fire bars* are seen at *f f*, etc., and the manner in which the top of the furnace is stayed is seen at *a a a*, etc.

106. Staying of the Furnace.—The staying of the furnace renders this end the strongest part of the boiler. The *flat* top is, of course, equally bad with the flat sides without the stay bolts, for all flat surfaces in a boiler are inherently weak. The top cannot be satisfactorily secured by stay bolts. The following plan is adopted:—Across the roof of the fire box are placed nine or ten *roof stays*, or *cross stays;* A B (fig. 3) is one of them; they are placed four inches from centre to centre, these roof stays are firmly bolted to the top of the boiler, as seen at *a a a.* The roof stays are further secured by *suspension stays*, or hanging stays *s s*, to the outer fire box in the manner shown very clearly in the figures at *c c*. Those in the upper figure are a little differently arranged to those in the lower, but the principle is the same in both. The roof stays are firmly bolted to the roof of the furnace, then suspension stays extend from the fire box roof stays to the top of the outside fire box.

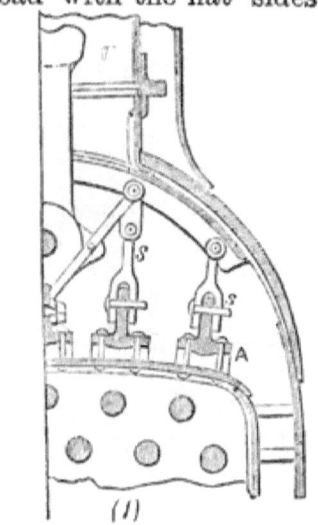

(1)

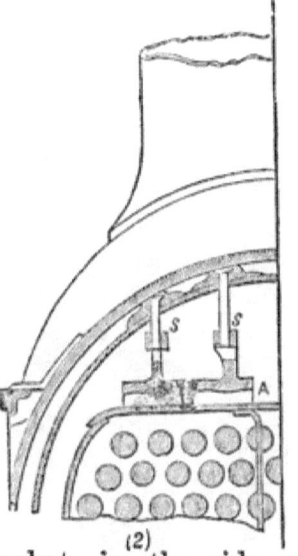

(2)

107. Fire Bars.—Fire bars of wrought-iron support the fire and separate the fire box from the ash pan. They are laid on a frame which rests on bolts or brackets in the side of the fire box. It is found that thin deep fire bars, laid close together, are much better adapted for the purpose of

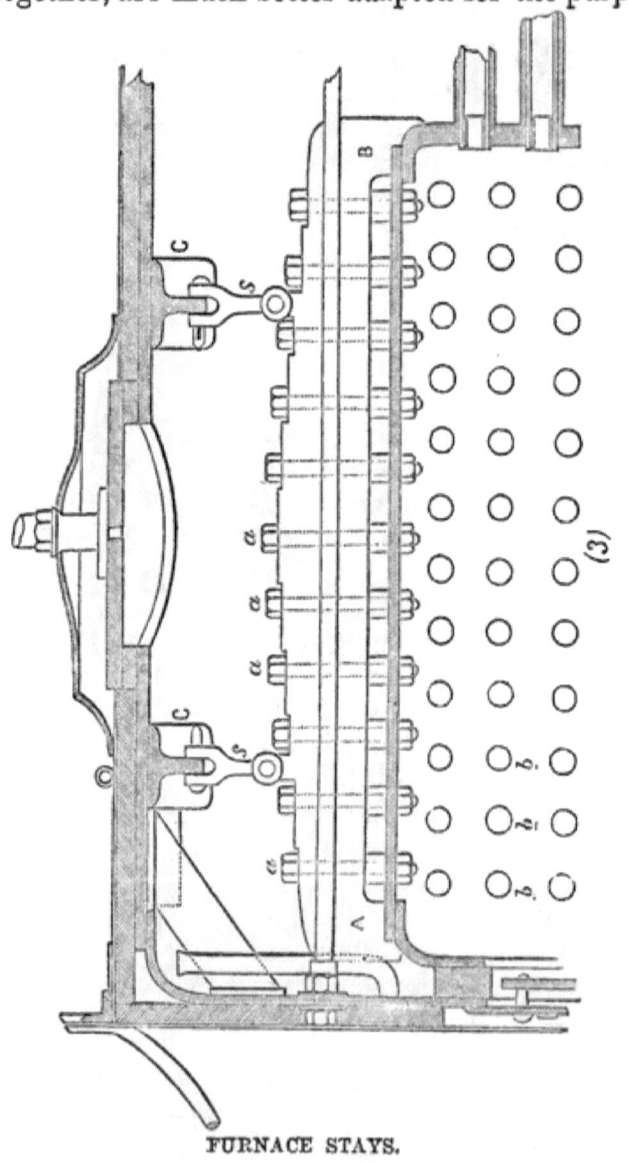

FURNACE STAYS.

a locomotive than larger ones. The fire bars, from the intense heat of the furnace, wear or burn away very rapidly. They are frequently bent—this arises from the softening of the iron from intense heat, when they drop, because they are not capable of sustaining the weight of the fire. Fire bars are about 4 inches deep, $\frac{3}{8}$ of an inch thick on the lower edge, and double that thickness at the upper, so that they are more widely separated on the side next the ash pan than on that on which the fire lies; they are so placed that the top of the bars are above the bottom of the water spaces by $2\frac{1}{2}$ or 3 inches. The fire bars are marked distinctly on Plate II., at fff, just above the ash pan A P.

108. Ash Pan.—The ash pan is placed directly under the fire bars, and is a simple wrought-iron tray about ten inches deep, the bottom being nine inches above the level of the rails. It must be carefully fitted and closed all round, so that the draught shall not be impeded, while the engine driver can use it as a damper to regulate the supply of air. Again, it should be so arranged that when the engine is running the air impinging against it shall be directed into the furnace. Its purpose is to prevent cinders and live coals from falling upon the line, for this, in early locomotion, caused several fires. There is another reason for it, as hinted above. When the engine is standing still, it is often important to stop the generation of steam, this is partly done by allowing as little air as possible to gain access to the furnace, hence the ash pan is made to fit tightly to the fire box on all sides; but the front side can be opened and closed at pleasure, like an ordinary damper, which is adjusted by a rod worked from the foot plate. When the engine is running rapidly with the damper opened, advantage is gained by the air rushing into the ash pan, and thence into the furnace. At sixty miles an hour, the pressure of air would be nine pounds per square foot, hence its advantage is at once apparent.

109. Smoke Box.—The smoke box is at the farther end of the engine to where the driver stands, or at the front of the engine exactly under the chimney. The heated air and products of combustion pass from the internal fire box through the tubes into the smoke box, and then are carried up the chimney. Access is given to it by means of a door, generally swinging on two hinges, which is kept fixed in its place as air-tight as possible, by means of bars, catches, and handles. Sometimes the door is in two parts, folding or overlapping in the middle, and closed by a bar, handles, and catches also. In the smoke box is placed the blast pipe, and the steam pipe runs down it to the cylinders at the bottom. Its use is to contain these, and to allow the tubes to be cleaned out, and to gather the soot, bits of coke, etc., that may be carried through the tubes. The smoke box is seen at S B.

110. Blast Pipe.—The discovery of the properties of the steam jet has been much disputed, some claiming it for one party, some for another. Its uses were fully understood before the year 1830.

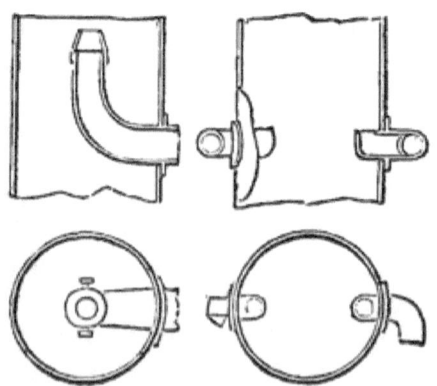

BLAST PIPE.

The above illustrations will show how the blast pipe was applied in the two cases of the "Royal George," on the Stockton and Darlington Railway, by Hackworth, and by Stephenson in 1827. The principle of the *blast pipe* has been previously explained. When the steam is introduced into the chimney, it causes a very powerful draught by rushing upwards and carrying with it the air, thus creating a partial vacuum, when the air rushes through the fire doors and bars to fill up the vaccum. In this act it carries a large amount of oxygen into the fire box, which assists in the more perfect combustion of the coke. The steam expands as it rushes out of the mouth of the blast pipe, and, filling the chimney like a plug, it not only drives all out before it, but drags with it the gases from the smoke box by mere contact. The degree of exhaustion in the chimney, or the vacuum, of a locomotive, is generally such as would support from 3 to 6 inches of water. The force of the blast greatly depends upon the amount of contraction given to the mouth of the blast pipe, as seen in the foregoing left hand figure. The contraction must not be carried too far, for it is evident that, if the steam cannot freely run out of the cylinder, a back pressure will be thrown on the piston. As there are two cylinders, the exhaust steam is led by a forked pipe, sometimes called a breeches pipe, toward the chimney, which joins immediately before it enters the funnel, where it stands up vertically in the centre. As the vacuum increases in the smoke box, there is an increase of blast pressure. This no doubt arises from the increase of speed, which means an increase in the rapidity at which one puff of blast succeeds another.

111. Safety Valves.—Two openings are made in the upper part of the boiler, which are covered by discs or valves. These valves are held down on their seats by levers; one arm, the shorter one, is secured directly to the boiler, while the other arm, the longer one, is held down by a stout spring balance, so screwed down that the valve can only rise when the pressure of the steam in the boiler becomes greater than the spring can resist. These valves are named safety valves, because by rising when the pressure of the steam exceeds the intended limit, they allow it to escape, and prevent any excessive accumulation of pressure whereby the safety of the boiler and persons around are endangered. The safety valve does not show how much the pressure of the steam may be below or above the proper limit; this is shown by the steam or pressure gauge. The safety valve of a locomotive should be placed as far from the dome as convenient, in order to prevent priming.

There are two generally fitted, one placed beyond the control of the driver and the other near him. They are kept in their places, one by a Salter's spring balance, and the other is held down directly by a spring secured to the top of the valve, and hence it has no lever; weights are inapplicable to the case of the locomotive, because they would jerk up and down with the vibration of the engine. The safety valves vary in size from $1\frac{1}{4}$ inches in diameter to 4 inches, but the general size is about 3 inches. Large safety valves are not so likely to set on their seats as smaller ones. The lever by which the Salter's spring balance presses the valve on to its seat is generally graduated, according to the area of the valve. If the valve be 10 inches in area, the lever is divided into 11 parts; the safety valve lever presses on the valve at the first division, leaving 10 divisions on the long arm and one on the short arm; thus the pressure per square inch on the safety valve is exhibited. They vary in shape in some engines; annular valves are used in which the steam escapes round the edges of two circles. The annexed figure illustrates a very good valve used by Mr Gooch. It is constructed somewhat on the principle of the steam indicator.

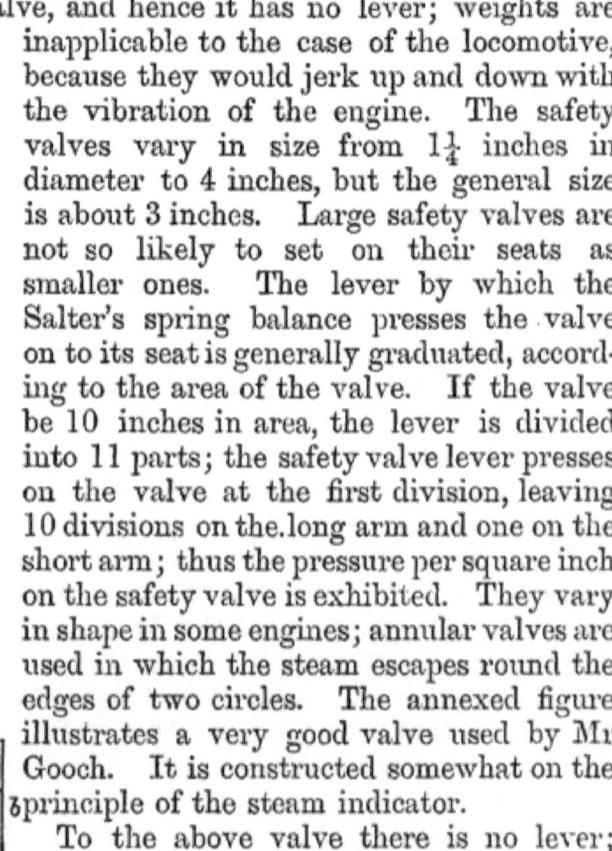

To the above valve there is no lever; the spring balance is placed on the top of the valve itself, which is $1\frac{3}{10}$ inches in diameter. The steam enters at S, when acting on $a\ a$ against the spring in the barrel B; the force of steam compresses it until it acts by allowing the steam to pass through *b*.

112. Chimney.—It is usual to term it a chimney, not a funnel. The height must not exceed fourteen feet above the level of the rails: they are made of wrought-iron, and proceed directly from the top of the smoke box, to which they are bolted. Their relative sectional area to that of the fire grate is about one-tenth, or they should properly be a little less in diameter than one of the two cylinders, which is considered a good proportion. Their draught does not depend upon their height; or, rather, the draught depending upon the rush of waste steam, it matters little what height they are, so long as they convey the steam, smoke, etc., away from the driver and fireman. A damper is generally provided, as seen in our figures, near the end of the blast pipe, and so arranged that the nozzle of the blast pipe passes through it. It consists of a disc of metal.

113. Dampers.—Besides the disc damper referred to above, placed across the chimney, the front of the ash pan is always so arranged as to act as a damper, by regulating the supply of fresh air to the fire. The most effectual dampers are those placed at the smoke box end of the tubes, consisting either of a perforated plate, with circular holes corresponding to the number and end of the tubes, which slides so to either completely close or leave open the ends of the tubes; or else it consists of thin strips of metal arranged and acting on the principle of the Venetian blind; by these the tubes can be left fully open or closed, or partly closed, so as to check the draught, according to the judgment of the driver.

114. Steam Dome and Prevention of Priming.—The position of the steam dome varies, but it is always bolted to its seating, which is riveted on the top of the boiler, sometimes immediately over the external fire box, and sometimes towards the middle of the barrel of the boiler. Within the steam dome is placed the end of the steam pipe, and it is here placed so that the steam shall enter it as far from the water as possible. It is sometimes known under the name of the **Separator,** because by the steam entering the steam pipe within the dome, a better chance is given for the spray produced by ebullition to separate from the steam—thus priming is prevented. Sometimes the safety valve is placed on the top of the steam dome, but this is considered an objectionable practice, as it should be as far away as possible from the steam pipe.

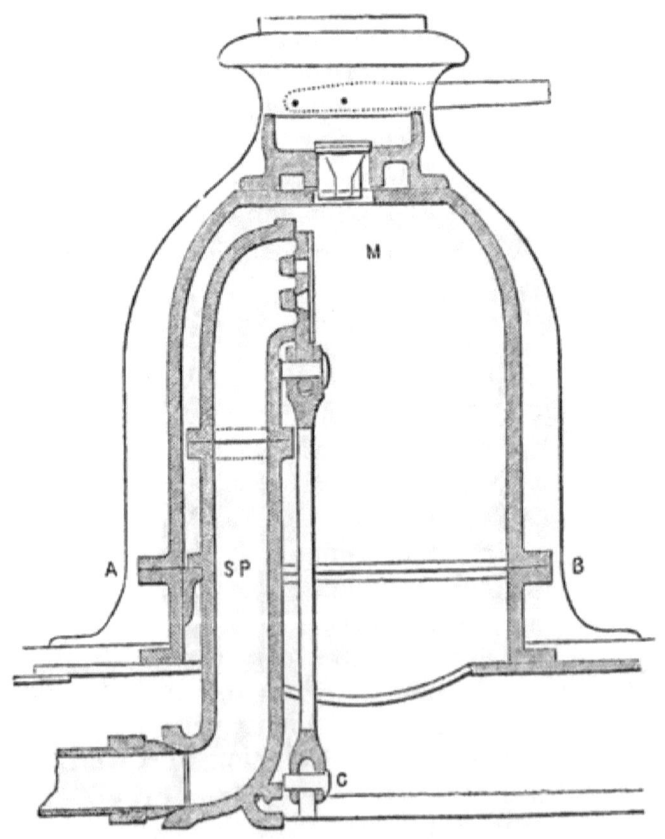

DOME AND STEAM REGULATOR.

115. Regulator, or Steam Regulator.—This contrivance is to regulate the admission of steam to the cylinders from the boiler. They are made in various forms, but are chiefly of two classes: (1) Those formed on the principle of a conical valve and seat; (2) those constructed like an ordinary locomotive slide valve.

C is a lever, or else an eccentric worked by *the regulating handle*, which is close to, or within easy reach of, the engine driver. This lever, or eccentric C, being moved, the slide M is brought down, and free exit is given to the steam in the boiler, so that it can readily pass down the steam pipe S P to the cylinder. Sometimes these valves are arranged precisely on the same plan as a ventilating grate in the floor of a building, where a very slight turn gives a large passage for air, in this case steam.

A baffle plate of brass, shown by the line A B in the figure, is fixed above the water line at the entrance to the steam dome—it is thoroughly perforated; as the steam runs towards the mouth of the steam pipe M S, it impinges against this perforated plate, and in rushing against the plate and passing through the holes, the water that has come away with the steam is knocked out—the whole arrangement is thus found effectually to prevent priming. Another mode of preventing priming is by placing the steam pipe as near the top of the boiler as possible, and allowing the steam to enter through holes in the top, before which are placed a smaller baffle plate; this has been already explained. The dome is bolted to its seat, which is riveted on to the top of the boiler or fire box—the former is the more preferable plan by far—and the joint is made steam tight, as explained under the next heading, Man Hole. Its form varies as much as its position, depending upon the taste of the maker, but the majority are either hemispherical or have hemispherical tops. It is usually worked out of one plate, with a spherical top, or finished with a dish cover of plate or cast-iron.

116. Man Hole.—The man hole is to give an entrance to the interior of the boiler. No special man hole is required, as the dome can be taken off, and admission thus gained to the boiler; but when fitted, it is frequently over the top of the fire box, or near the chimney, or on the dome seating. Near the fire box is the best place, as the stays can easily be reached. It is about 15 inches in diameter, sufficiently large to admit a man's body. The door of the man hole must be attached with a steam-tight joint to the top of the boiler; it is rendered steam-tight in the ordinary way, by the use of canvas and red lead. Sometimes the molecular force of expansion is made to render the joint steam-tight, thus:—Soft copper wire is laid on the joint, then the cover is brought down on to it and screwed up as tightly as possible, when the steam is up the heat causes the copper wire to expand; the greater the heat, which, in this case, may represent pressure of steam to escape, the greater the expansion of the copper, and the more steam-tight the joint. It is made with a necking formed of thicker metal than that of the boiler, and flanged to join it The upper flange is planed to receive the cover or dome.

117. **Steam Whistle.**—The steam whistle is a device attached to locomotives for giving warning that the train is approaching, moving, etc. It mainly consists of a pipe fastened into the top or end of the boiler, with a cock within easy reach of the engine driver. When the steam is turned on, it issues violently out of a circular opening and strikes the rim of a bell-shaped piece of brass (its edge being placed exactly over the circular opening), with sufficient force to make the whistle heard at a very long distance. The principle is simply this:—When the handle H is turned, the steam coming from the boiler passes up S P, and out all round the edge of ss through the circular opening cc, then impinging with great force upon the edge of $s's'$ it sets it vibrating, the vibrations communicate their motion to the air, and mould it into a series of sonorous waves, giving us a high note of so shrill a pitch that it can be heard at a very considerable distance.

There are generally two whistles—the shrill one for ordinary purpose, and a deep-toned one to attract the guard's attention. It is usual now to arrange the guard's whistle, so that both the driver and guard can sound it. The cord that runs along from one carriage to another is in connection with this whistle, and if the passenger pull this cord he will sound the deep-toned whistle.

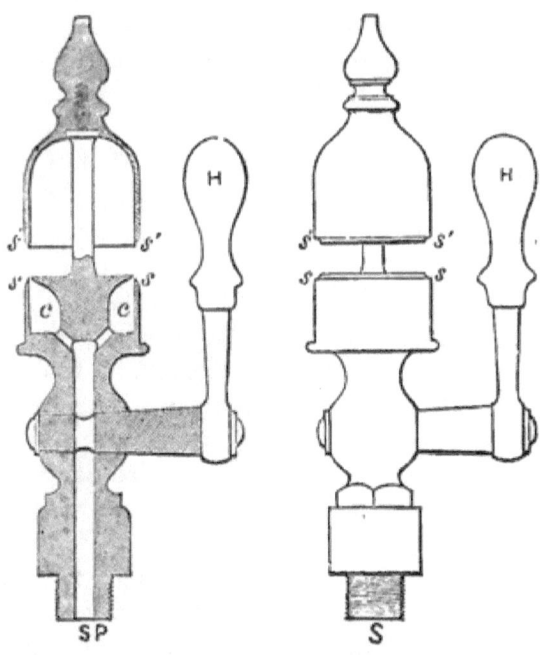

STEAM WHISTLE.

118. Salter's Spring Balance is used especially in locomotives to exhibit the pressure of steam. Its principle is a steel spring, well tightened, which, according to the pressure of steam, extends after the manner of the spring steel yards used in public by our rag and bone merchants; or else the increased pressure of steam acts against the spring.

An adaptation of the spring balance to safety valve, is shown by the following figure, where A is screwed into the boiler, or into a pipe in free communication with the steam, so that steam can enter the cylindrical body B; if we suppose the dotted lines at A are a piston, it will act against it to drive it down, which the pressure of the spring will not allow it to do until it overcomes its resistance. The greater the force of the steam the more will the spring be compressed, and the more of the graduated part be shown. Acting on this principle, it is evident that, if it be properly graduated, the pressure of steam in the boiler will be correctly indicated by the scale. When used to keep down a safety valve, it is evident the valve must be able to rise considerably on its seat before steam can escape; this is the exceptional arrangement, as Salter's spring balance is used in a simpler manner for a *pressure* gauge, and not to keep the safety valve on the seat.

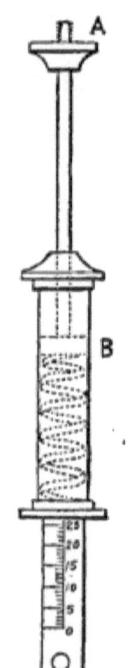

SPRING BALANCE.

119. Bourdon's Gauge.—This gauge is produced in many shapes—we give one of the most portable and convenient in the figure on the next page. A B is a circular plate, fitting steam tight in *s*, but still readily moving with the least pressure. *s* is in free communication with the boiler, by way of E; therefore the pressure of steam below will cause the plate to ascend, when the rod *r* will move the lever *a b* on its centre *b*, and with it the rack *c d*, which moves the pinion *p* from right to left, and with it the pointer P, which will indicate the number of pounds pressure in the boiler on the arc.

The use of gauges, it will be gathered from what precedes, is (1) to tell *accurately* the pressure of steam in boilers when water is hotter than 100°C.; (2) to indicate the *variation* in the pressure of steam from time to time. When we consider how much depends upon a knowledge of these facts, the following instance of, to say the least, carelessness and thoughtlessness will astonish us:—Out of 52 gauges tested for the Royal Agricultural Society, upon the occasion of their exhibition being held at Manchester, only nine were correct. If this be a fair average, the deplorable fact comes to light that only 17·3 per cent. of the gauges in common use give correct indications of the state of the boiler pressure.

120. Vacuum Gauge.—The same figure will illustrate the vacuum gauge and its principle. This gauge is to show the state of the vacuum in the condenser, and so is an appendage to the condenser and not to the boiler. E is fitted into the condenser. If A B be air tight, there being a vacuum in the condenser, when the cock V is opened the piston will *descend* by reason of the pressure of air above it. If the pointer be directed to a particular point when the air is acting freely on both sides of the piston A B, then, as the vacuum increases in the condenser, the pointer will move right to left. When the gauge is used to show a vacuum, the graduation only extends from 1 lb. to 15 lbs.

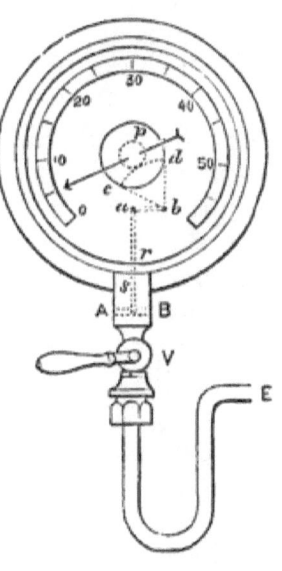

BOURDON'S GAUGE.

The teacher must accustom his pupils to draw the figure clearly, pointing out the difference of action when used as a vacuum gauge and as a steam pressure gauge.

121. Sector.—The *sector* is in the form of a sector of a circle, and is an adjunct to the link motion. In it is a series of notches to hold the reversing handle. When the locomotive is started, the handle is dropped into full throw, or into the farthest notch, when all the strength of the steam is at once given to the piston. There are about five notches from the centre down, or ten altogether. The nearer the reversing handle is to the centre of the sector the less steam is used.

CHAPTER VII.

THE WATER FOR A LOCOMOTIVE

Water Tanks—Water Cranes—Feed Pump—Giffard's Injector—Gauge Cock — Glass Water Gauge — Screw Plugs — Scum Cocks—Blow-Off Cocks—Heating Cocks.

122. Water. — The boiler of a locomotive engine is filled so that the water stands a few inches above the top of the fire box. It is admitted to the boiler by means of pumps and ball valves, or by Giffard's injector.

123. Water Tanks.—Walls, or small buildings of substantial masonry, supporting a large tank for water, are generally seen by the side of a railway station; they supply the engine with water. Water tanks are usually rectangular, from five to nine feet deep. They are at the bottom, at the least, twelve feet above the level of the rails, so that there is constantly a sufficient pressure of water to fill the *tender* quickly. Tanks are either filled by allowing the water to run into them from a higher level, or by pumping up the water by means of an engine from a lower level. This is the general case when the engine, boiler, pumps, etc., are housed under the tank. The water tanks are made of boiler plate-iron, and supported by cast-iron beams running in a row under each seam of the tank. They are also made of cast-iron, supported by cast-iron beams across the tank from side to side. The engines preferred for the purpose are *vertical*, and the pumps double-acting.

125. Water Crane. — The water is drawn from the tank at the bottom, and passes through a cast-iron pipe to the *water crane*. It is allowed to pass into the mouth of this cast-iron pipe by a valve which has the fulcrum of its lever on the side of the tank; the valve is lifted for the discharge of water by a chain hanging outside within reach.

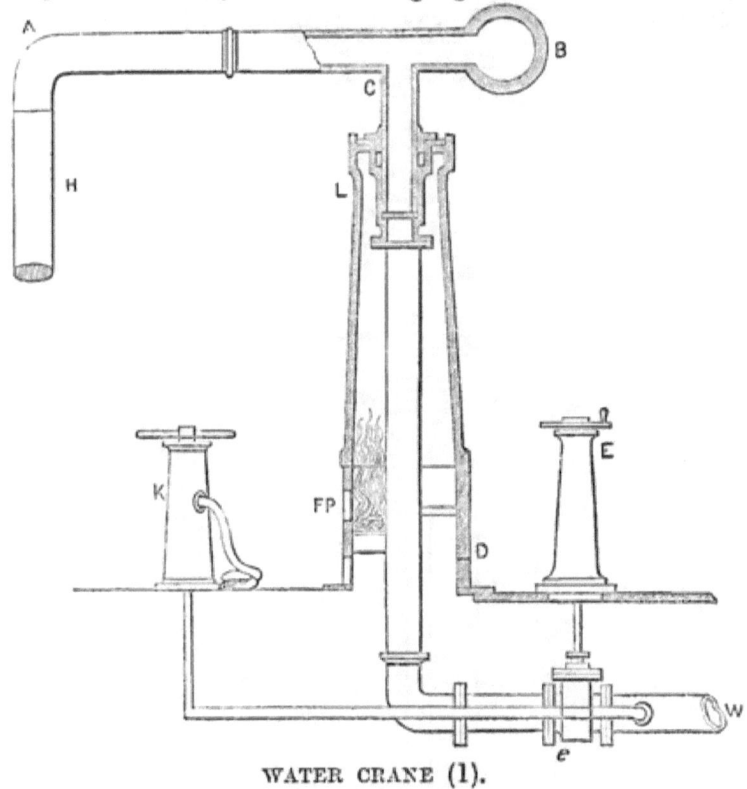

WATER CRANE (1).

Fig. 1 is a water crane of the usual construction. A B is the *swing pipe*, balanced on a vertical pivot at C, within the cast-iron column C D; it will, therefore, swing round into any position convenient for filling the *tender with water*. H is a leather hose at the extremity of the swing pipe for the convenience of the engineman. E is a shut-off screw valve, to allow the water to pass up the column D C when the handle is turned, and to stop the supply when sufficient has been delivered into the tender; the valve *e* is screwed up when water can pass from W. It will be seen that its action is exceedingly simple. Water passing from the tank, by way of the pipe W, is allowed

116   STEAM.

to run through the valve *e* by turning the handle at the top of the shut-off screw E, it then goes upwards through D C, along C A, and into the tender by H. B is a weight to counterpoise that of A C, so that no undue strain comes

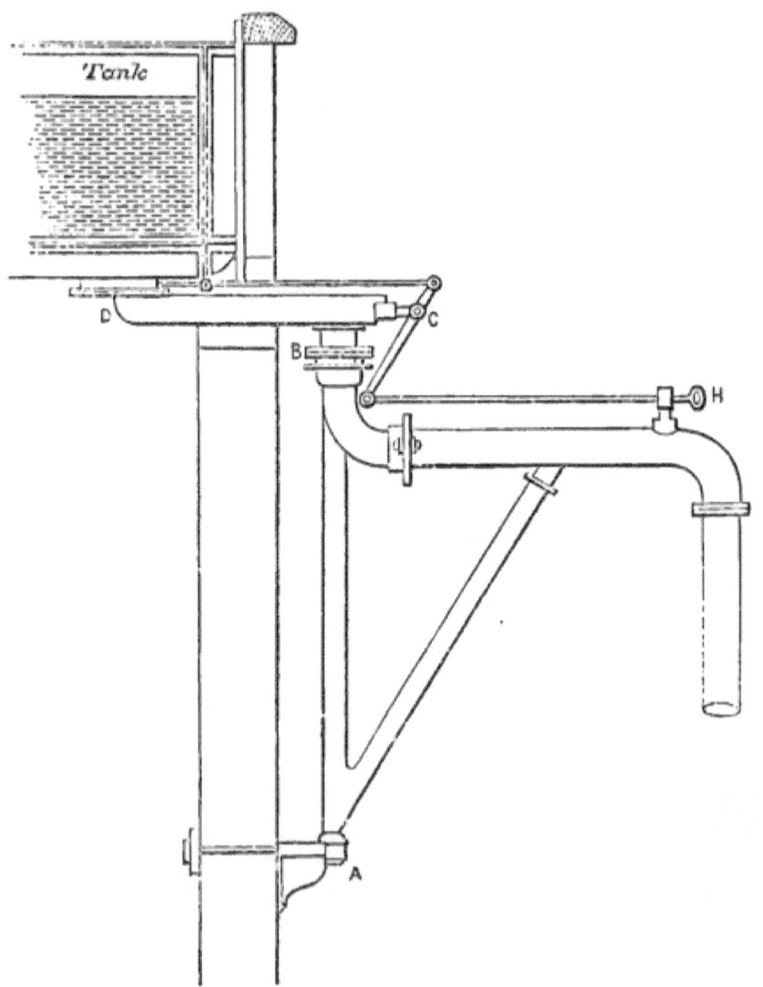

WALL WATER CRANE (2).

on the vertical pivot C; also, by its momentum, it assists in turning the arm A C. It will be observed at C that the pivot has a brass bearing, which has to be fitted with considerable care. F P is a fire place, so that in winter the column can be warmed and the water unfrozen, or prevented from freezing; the products of combustion pass out through a number of small apertures provided for the purpose at L. K is a pillar fountain, from which water can be taken, by turning the handle at the top, for cleansing and other purposes; an hose can be attached to it for the convenience of watering and cleaning.

The *wall water crane* is simpler in its details than the one just described, but not always so well adapted for its purpose, as it makes no provision for the extreme cold of winter. It swings at the bottom A (fig. 2) on a bracket bolted to the wall, and at the top B it is supported by the supply pipe D C, into which it is pivoted. The engine driver pulls the handle H, when, by means of lever C, a sluice valve is pushed back within D, when the water runs along the supply pipe D C, and into the swing pipe, as before, to the tender through the leather hose at the extremity of the supply pipe. The tank is seen in its proper position.

124. Ramsbottom's Self-filling Tender. — It being necessary to run express engines long distances without stopping, and as the requisite supply of water is very great, from 1800 to 2400 gallons (8 or 10 tons), a method to prevent starting with such a dead weight was put in practice by Mr. Ramsbottom. It was necessary to commence the journey with 11 tons of water for fear of accidents. He put down a pair of water troughs on a level part of the line near Conway. They were of cast-iron, 18 inches wide, seven inches deep, and 441 yards long, with a further length of 16 yards at each end, rising one in a hundred. They are so placed that water of about 5 inches deep stands at about the level of the rail. A tender of the ordinary construction capable of holding about 1500 gallons is provided with a rising water pipe, the lower end being fitted with an adjustable scoop ten inches wide at the mouth. It is clearly seen between the two wheels,

118 STEAM.

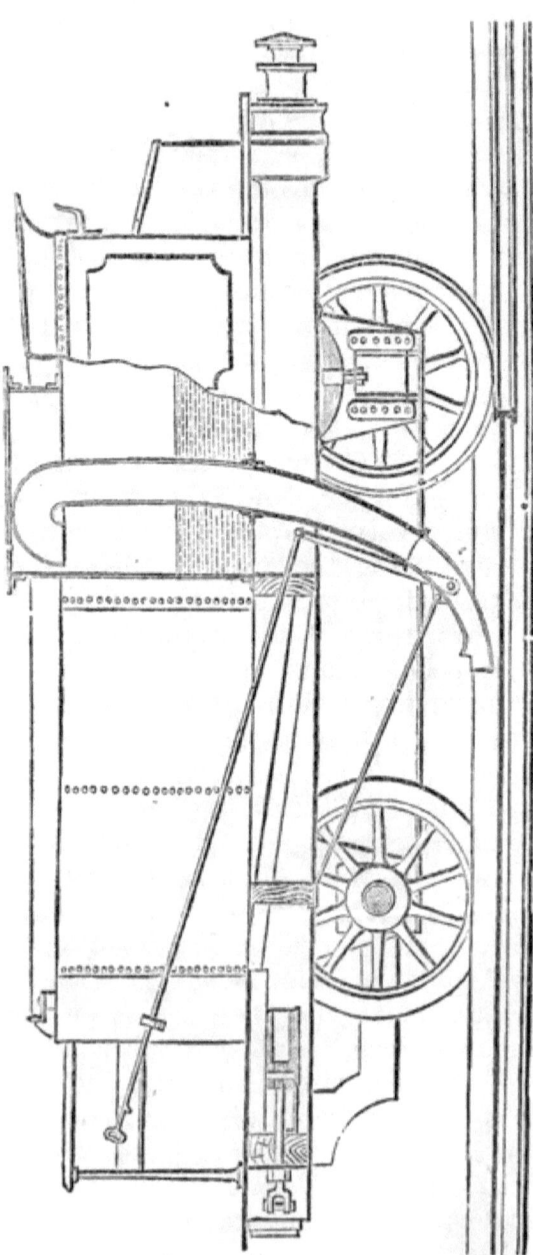

RAMSBOTTOM'S SELF-FILLING TENDER.

with the rod for lowering it at the proper moment, in the annexed figure. The scoop meets the water, and for the purpose of filling the tender, we may suppose the scoop at rest and the water in motion at the velocity of the train. Whichever way we consider the matter, the water must rush up the pipe, and if allowed to do so would rise to the same height that a body must fall to acquire the velocity of the train. At 22 miles an hour, this would be 16 feet high, at 60 miles an hour, it would be 121 feet. So we see that between, or even below, these velocities, there is ample force to fill the tender. The scoop dips two inches under the surface of the water, and at 22 miles per hour 1000 gallons are raised into the tender in less than half a minute. The delivery is practically the same at higher speeds, for although the velocity of the entering water is then greater, the time during which the apparatus acts is diminished in the same proportion.

125. Feed Pumps.—The feed is either supplied by *Giffard's injectors*, fixed to the fire box, and of which a description will be given, or by an ordinary double-acting pump worked off the crosshead of the piston-rod, or from one of the eccentrics on the crank axle. When the former method is adopted, the ram is about $1\frac{7}{8}$ or 2 inches in diameter; but in the latter arrangement the ram must necessarily be of greater diameter, about 4 inches, as the stroke is so much shorter.

The water is kept in the tender T. The handle at h is turned, when the plug p is lifted, and the water runs down $p\,b\,c$ by gravity. At b is a ball and socket joint, so that the pipe $b\,c$ (this part is generally called "bags") is capable of a slight vertical and lateral motion. From d to e is a telescopic joint, which admits of a longitudinal motion in and out. It is thus that all the motions of the train are provided for, and that the joint is rendered water tight. At c it is screwed on the pipe leading to the engine and boiler. This tube leads the water to W in the figure on p. 121, which gives us two views of the feed pump. p is the plunger, a side view of which is shown

120　　　　　　　　　STEAM.

at A. The eye of the plunger rod is fastened to the crosshead of the piston, but sometimes to the back of the eccen-

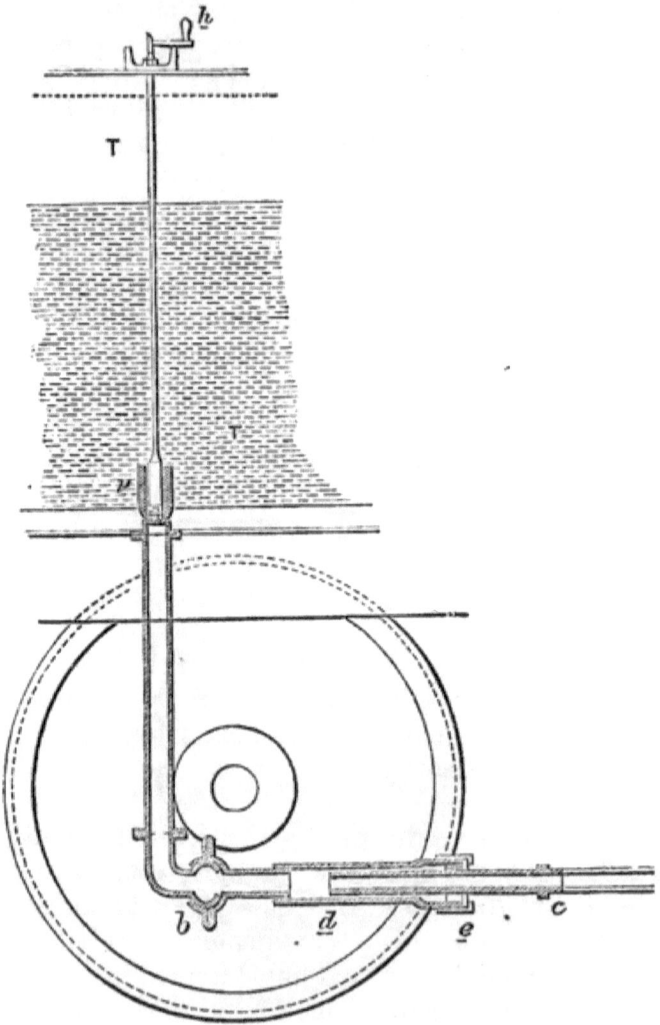

BALL AND TELESCOPE VALVE ON TENDER.

tric to the eye at G, as seen in the figure of Stephenson's Link Motion. The plunger is very small, not more than 2 inches in diameter. As it moves to the right a vacuum is left behind, and the water rises through the valve v; next, as it comes back, the water is forced along the delivery pipe from v' to v'', through the ball valves v' and v'', into the boiler at B. The object of the third valve at v'' is to prevent the pressure of the steam from forcing the

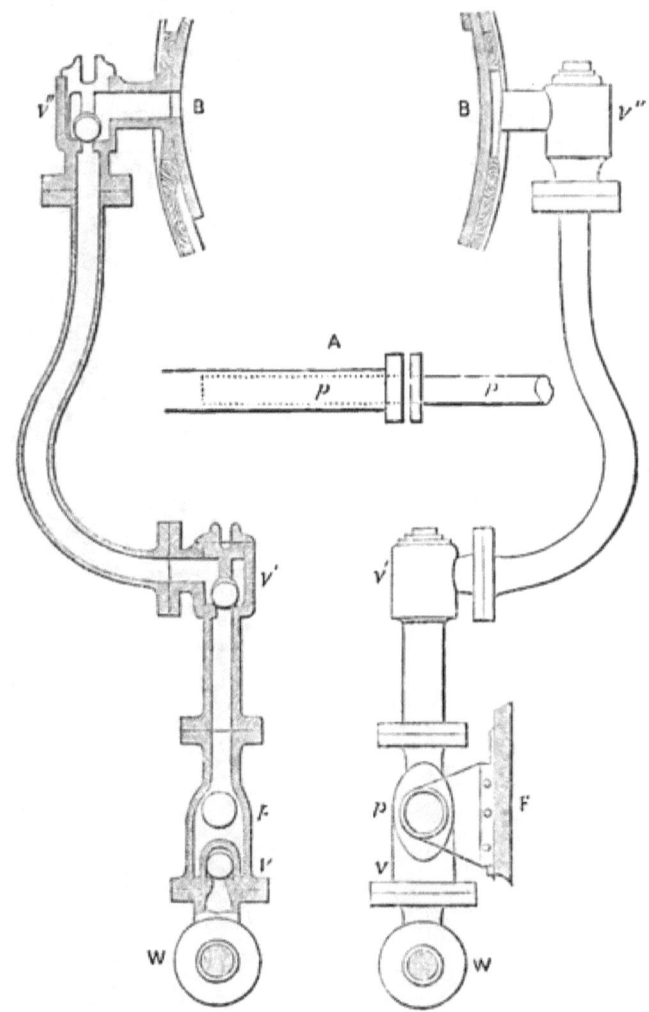

FEED PUMP.

water back upon the other valves. The lift of the valves is very small, not more than a $\frac{1}{4}$ to $\frac{5}{16}$ of an inch. Above each valve is a guard to keep it down to its seat; for, if allowed to rise too high, the force of concussion would be sufficiently great to destroy the valve seating. When no feed is required the water is shut off at the tender. These feed pumps only work when the engine is moving. Sometimes it may be noticed that engines are running backwards and forwards a short distance near a railway station. It is that water may be pumped into the boiler. When Giffard's injector is fitted, there is no necessity for this. It was a custom to fit a small donkey pump for the purpose of forcing water into the boiler when the engine was stationary. The capacity of the pump, *i.e.*, the area of the plunger or ram, multiplied by the length of the stroke, should be from $\frac{1}{70}$ to $\frac{1}{80}$ of the area of the cylinder. Each pump or injector should be capable, singly, of keeping up the feed. Two are fitted in case one should be disabled.

126. Gauge Cocks.—When the boiler is first filled with water, it is made to stand a few inches above the fire box. In order to know when the water is at the proper height in the boiler, there are fixed in the back or side of the fire box two brass gauge cocks. One is a few inches above, and the other as much below, the proper level of the water in the boiler. The cocks are connected with a glass tube, the whole forming the **Glass Water Gauge**. Both cocks are kept in communication with the boiler, so that the water can freely pass through the bottom cock into the glass tube, and the steam as freely through the top one. The water within the gauge has thus the same level as that in

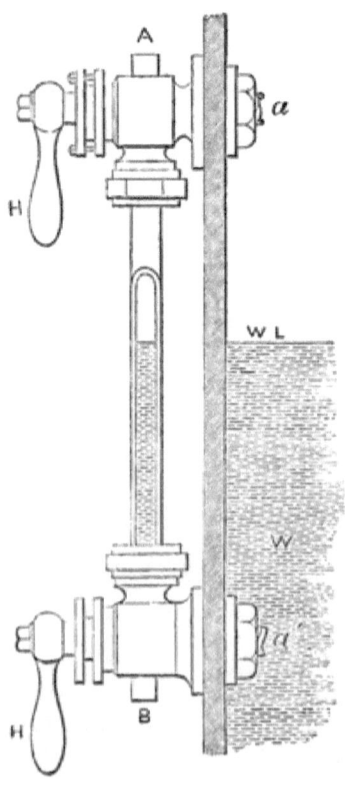

GLASS WATER GAUGE.

the boiler, and the driver has only to look at the glass to see the height of the water in the boiler. When the feed pumps are at work, he watches it till there is a sufficient supply in the boiler, and afterwards he has to notice that it does not get too low through the evaporation of the water. In addition to this there are fitted three *gauge cocks* at the back or side of the fire box at different heights, between the extreme limits admissible for the water level. By trying these cocks successively, the engineman can judge, according as steam or water issue from them, at what height the water stands in the boiler.

W is the water in the boiler, W L is the water level. The water passing from the boiler enters at a', and stands the same height in A B, the glass water gauge. The handles H and H, when turned, allow water or steam to issue from the boiler, and clean the gauge out. It is the duty of the engineman to turn these handles now and then, for fear the gauge may be choked at a, or a', or at B and A.

127. Screw Plugs.—To facilitate the washing out of the boiler, a screw plug, about 2 inches in diameter and slightly tapered, should be fitted at each corner of the fire box, with as large square heads as the plugs will admit, to bear the strain of the screw key. The plugs should be of hard brass, and threads cut to a fine pitch, to give them a good hold on the metal; sometimes a lining plate is inserted at the corner to increase the hold of the plug, and reduce the liability to leakage.

128. Scum Cock.—This cock is fixed on the back of the fire box at the ordinary water level, with $1\frac{1}{2}$ inch copper pipe, carried down below the foot plate, to draw off the impurities which rise to the surface of the water, and which, whilst there, frequently cause the boiler to prime.

129. Blow-off Cock is also fixed at the back or side of the fire box, but at the level of, or as near as practicable to, the ring at the bottom of the water space between the internal and external fire boxes, and is for the purpose of blowing the water out of the boiler when required. There are two other cocks fitted to all locomotive boilers, viz., the blower and the warming cock; the former being connected by a pipe with the chimney, for the purpose of getting up the steam rapidly, the latter for warming the feed water in the tender. It is generally opened while the engine is stationary, when by suitable pipes the steam passes to the tender, where it heats the water instead of blowing off to waste. This practice was adopted in the very earliest days of locomotive engineering. It is also a common practice to heat the water by other methods before it enters the boilers; in fact, this should always be done if possible.

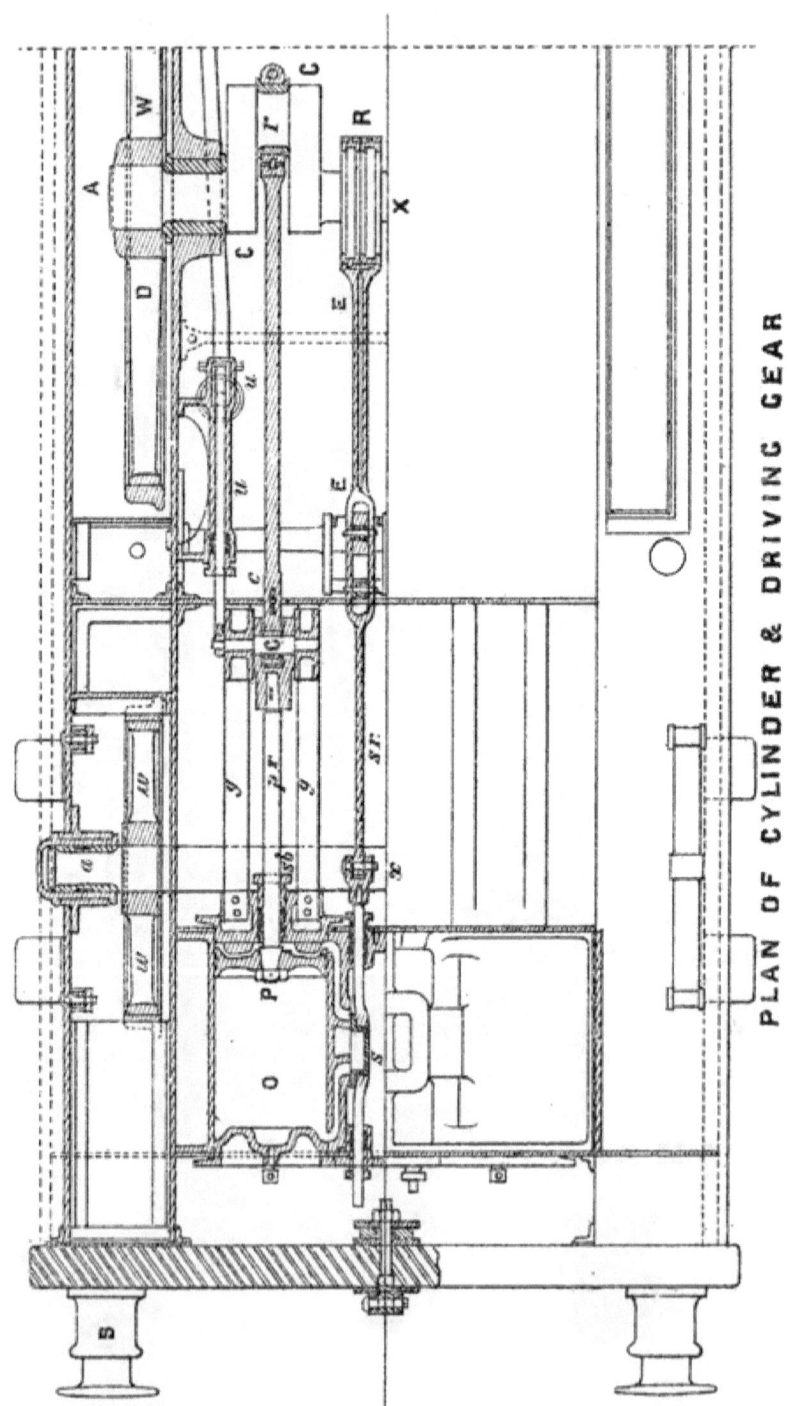

PLATE III.

PLAN OF CYLINDER & DRIVING GEAR

CHAPTER VIII.

The Cylinders—Water Cocks—Grease Cocks—Piston and Piston-Rod—Connecting Rod and Crank—Coupling Rod—Strap, Gib, and Cutter—Sector—Driving Wheel Tire—Counter-weight to Wheels—Sand Cocks—Axle Boxes—Springs, Buffers, and Buffer Springs—Brakes—Draw Bar.

131. The Cylinders of locomotives are generally placed immediately beneath the smoke box, where all condensation from external cold is entirely prevented. Sometimes they are fixed on the outside of the engine, such engines receiving the name of Outside Cylinder Engines. In early locomotives the cylinder was placed vertically. The horizontal cylinder was finally adopted about 1830. It is unnecessary to enter into the details of the cylinder. The student is referred to what has been already said concerning those of the land and marine engines generally, as the arrangement is the same. Of course they are made of good hard cast-iron. Sometimes, from the weight and friction of the piston, there is a tendency to groove. They are generally constructed so that both the top and bottom of the cylinder may be removed. The piston-rod works through a stuffing box and gland in the ordinary manner. It is always usual to allow one-quarter of an inch, or less, clearance at both ends of the stroke.

In Plate III., the cylinder is seen at O with its piston P. The piston-rod is $p\,r$, and two guide blocks are at G, which move backwards and forwards between the guide or motion bars $g\,g$. The piston crosshead is also at G. Into the guide blocks comes the end of the connecting rod $c\,r$. C C is the crank moved round by the connecting rod, and carrying with it the axle A X, and with it the driving wheel D W. R is the eccentric, and E E the eccentric rod working the slide rod $s\,r$, which in its turn gives the reciprocating rectilinear motion to the slide s. The slide s is seen in front of the ports, the bottom port being open to the exhaust and the upper to the steam. The piston is just going to commence its stroke to the left. The manner in which the connecting rod is attached to the crosshead of the piston and to the crank is explained by the illustration. $a\,x$ is the axle of an ordinary leading wheel $w\,w$; the part marked a is the journal.

132. Water Cocks, Drain Cocks, Relief Cocks, or Cylinder Pet Cocks.—Two drain cocks are fixed to each cylinder, one at each end, and at the lowest part, to relieve the cylinder from any water that may arise from condensation of steam or priming. They should be opened just before starting, after the engine has been standing still, to get rid of any water that may have become condensed while waiting. They are worked by rods and levers from the footplate. Sometimes, often after repairs, the water is greasy, and until the grease is properly got rid of, the engine will often prime, hence the value of these relief cocks.

133. Grease Cocks.—A grease cock is fixed on to each cylinder; it communicates with the slide valve and lubricates it; part of the tallow, as the slide moves backwards and forwards, enters the cylinder and lubricates it also. It is generally fixed on the valve-jacket, so that the slide valve and cylinder are lubricated as well.

134. The Piston and Piston-Rod.—The crosshead is the part to which the farther end of the piston-rod is fitted, to this also is attached the connecting rod, the crossheads move in guides or between motion bars, which are two or four parallel bars.

Pistons for locomotives are fitted and packed in many various ways. The piston-rods are made of steel or iron, while the piston itself is of cast-iron or brass; brass is the better substance, because it is lighter and does not so readily break; some makers forge the rod and piston in one piece. The top of the piston-rod is fastened by a cutter into a socket with jaws; G is the socket, the jaws are a little above G to the right and left; the whole is named the piston cap. Between and into the jaws comes the small end of the piston-rod $p\,r$, which is kept in its place by the pin of the crosshead; the two ends of this pin are fastened into two blocks, which move in guides or motion bars, to preserve the parallelism of the piston-rod. The pin of the crosshead is seen running under G, while the guides are marked g and g, and the two guide blocks may be observed to the right and left of, or above and below, G at the end of the guide bars. The piston-rod works steam tight through the cylinder cover; between P and $s\,b$ is a short tube cast on the cylinder, with an opening a little larger than the diameter of the piston-rod, this is called the stuffing box, the gland is the part close to $s\,b$. The piston-rod being in its place, the stuffing box is first filled with hemp soaked in melted tallow, or else with other packing; the gland is then brought down on to it and screwed forcibly against the packing, so as to press it tightly against the piston-rod. Whenever any sliding rod has to work into a space filled with steam or with water under pressure, a similar method is adopted to prevent any escape at the side of the rod.

"The maximum economical speed of the piston has not been ascertained, but it appears that with a high speed of piston, small driving wheels and light engines are preferable to the very large ones which are now frequently seen. Small wheeled engines have been found to start a train more rapidly and to draw it with greater regularity of motion than engines with $6\frac{1}{2}$ feet to 8 feet driving wheels."

135. The Connecting Rod and Crank.—By the intervention of *the connecting rod and crank*, the rectilinear motion of the piston is converted into a circular motion,

The connecting rod is c r, crank C C, and axle A X (Plate III.), which move the driving wheel D W. The crank, or rather the cranks—for there are two, as there are two cylinders and pistons—are forged on the axle of the driving wheels. The cranks are placed at right angles to each other; only one is seen in the figure, the other half is precisely similar to A X, but the crank is at right angles to C C. In our illustration, C C is lying horizontally, so that when the piston P attempts to move to the left, it will only pull the crank in a straight line, as it were, and cannot move it, hence we see the necessity for two cranks; the one not shown being at right angles to C C, is just in that position where the piston will have the greatest effect upon it; hence the driving wheel can be moved, which could not happen if the engine stopped exactly as seen in the figure, and there were only one crank. Such an axle as we have here is called a cranked axle, and is made of wrought-iron or steel. It must be understood from what precedes, that when one piston is at the end of its stroke the other is in the middle.

131. Coupling Rod.—A coupling rod is very similar in its form to a connecting rod, but it is not so large or heavy. Its use is for coupling the driving wheels to the leading or trailing wheels, or both, when of course the wheels must all be of the same diameter, as in the case

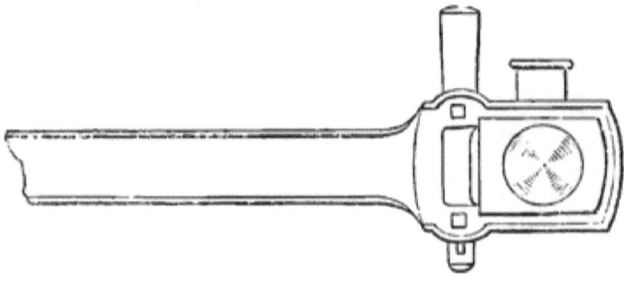

of goods engines. They are attached to cranks fixed on the outer ends of the axles, or else to crank pins inserted in the arms of the wheels—the former method applies to engines with outside bearings, and the latter to those with inside bearings. They are always outside the wheels. Generally they are made with ends forged in one piece, and the cutters so arranged as to preserve their length constant as the bushes wear. An oil cup is shown in the figure; it is forged on and has a small tube in the centre, in which to insert the wick to lubricate the bearing.

137. Driving Wheel.—The wheels attached to the crank axle are called the driving wheels; the front pair of wheels of the engines are called the leading wheels; and the hind wheels, or those close to the fire-box, the trailing wheels. They are nearly always made of wrought-iron, and are kept upon the rails by a flange formed on the tire. The driving wheels in passenger engines are always made large to increase the speed, and the power of the engine must be increased in the same ratio; but in goods engines they are not so large, and consist of four or more coupled together by coupling rods. The object in employing coupled driving wheels is simply to distribute the great weight necessary for adhesion where great tractive force is to be exerted at moderate speed, such as with a goods engine.

138. The Tire is a distict part of the wheel, composed of a ring of metal, either wrought-iron or steel, which is shrunk on to the wheel, and further secured to the rim by bolts or rivets.

Wheels are made upon various systems, the object of all being to give strength to the tires and prevent wear. The tires are not cylindrical, *i.e.*, they have not the inner and outer edge both the same diameter, but are made slightly conical, which plan keeps the carriage in the centre of the railway, and the flanges do not come in contact with the rails unless under exceptional circumstances; in fact, conical wheels have a self-adjusting action, which preserves the carriages in their proper position on the rails. Again, if the wheels be thrown into such a position that one flange is close against the outer or large curved rail, the wheels being conical, a larger circumference of the outer wheel will move on the rail than on the inner wheel (for we must recollect that the wheel only rests on *one* point), consequently the difference will quickly restore the carriage to its proper position.

139. Counterweights to Wheels.—The momentum of the piston-rod, guide blocks, connecting rod, etc., is very great; this has to be counterbalanced by the application of a weight to the wheel. These weights are put into the rim of the wheels between the spokes. If the student will notice the driving wheels of a locomotive, he will see the balance weight partially filling up the space between three or four of the spokes. This weight depends upon the speed at which the engine is intended to run, and the weight of the moving parts; with the engine-maker this is a matter of nice calculation. Seven-eighths of the whole disturbing weight is allowed with outside cylinder engines, and for inside cylinder three-fourths. Counterbalancing is done to give the engine greater stability on the rails. It is said that engines, without counterbalancing, will not attain the speed they will when counterbalanced, the resistance being greater. They must be sufficiently heavy, not only to balance the crank and connecting rod, but the piston and its appendages.

140. Sand Cock.—To every engine there is a small sand box, fitted either on the top of the tank, in front of the engine on the buffer beam, or by the side of the foot-plate. In connection with it is a small pipe, from $1\frac{1}{4}$ to 2 inches in diameter, leading to within two inches of the rail in front of both driving wheels, or in front of the whole if connected by coupling rods. The cocks are opened in slippery or damp weather, when the engine is starting, to assist the wheels in biting the rails, so that they may not run round without giving motion to the engine. The engine being fairly started they are closed. Whenever the wheels begin to slip, the cocks are opened till the nuisance is abated; and are, as occasion may require, brought into use on inclines.

141. Axle Boxes.—The wheels are fixed securely upon their axles, which revolve in boxes, upon which the weight of the boiler and machinery is carried through stout springs. The axle boxes can rise and fall freely, so far as the springs will permit. The axle boxes are guided vertically by suitable guides or *axle guards*. The part of the axle which revolves in contact with the axle box is called the journal. When the journals are inside the wheels they are called inside bearings, and when outside the wheels outside bearings.

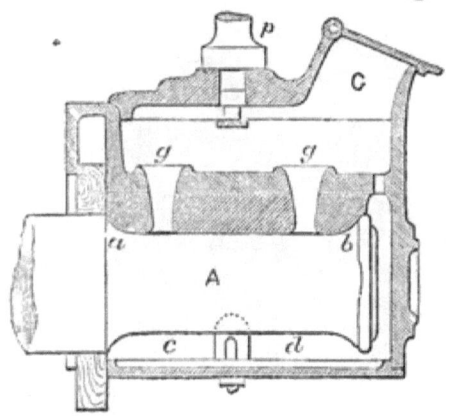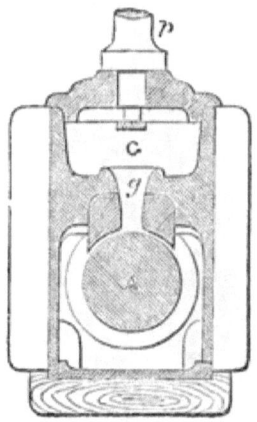

AXLE BOX.

A is the journal, the whole weight rests on the spring, of which *p* is the spring pin, therefore the weight of the engine rests on the top of the axle (and wheel) from *a* to *b*; *c d* is hollow, although sometimes a sponge, or some cotton waste, is laid in to soak up the oil or grease. In the cross section it is seen more clearly where the weight rests upon the axle.

142. Springs.—The weight of the engine, boiler, etc., is sustained by springs resting upon the axle boxes. They are formed of steel plates, from three-eighths to half-an-inch in thickness, of a number proportioned to the weight they have to carry. Each spring of the driving axle has often to carry from four to six tons. The plates are connected at the centre, and slide on each other at their ex-

132 STEAM.

tremities. If we examine the spring A, we shall notice a rod proceeding from the centre of the spring s' to the top of the axle box at d. The middle of the spring thus

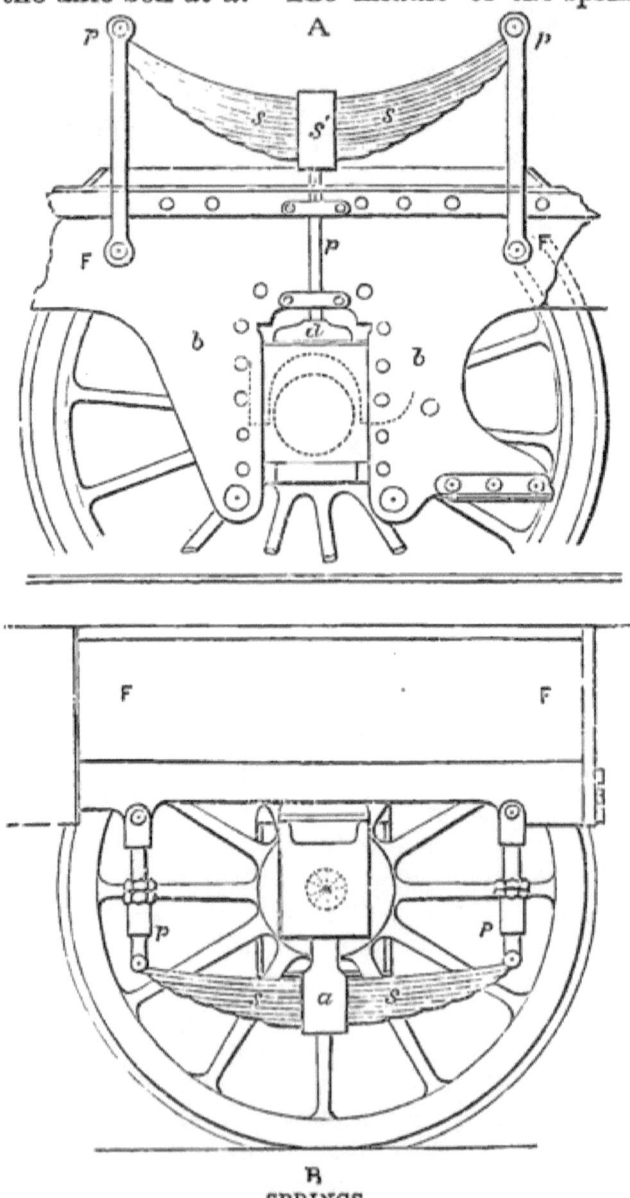

SPRINGS.

rests upon the axle box. At p and p are two eyes, the ends of the spring pass into the jaws of a bridle at p and p, and through them passes a pin to keep the spring firm at p and p. Sometimes, as in figure B, the springs are placed below the framing, when the weight of the engine is made to rest upon the ends of the springs. In figures A and B the weight of the engine is carried by the springs $s\ s$, the framing F F resting on the spring pins $p\ p$, the springs then bear up the weights. They are fastened to the axle box in figure B by means of a pin passing through the eye of a strap a round the middle of the spring. In the upper figure $b\ b$ are the horn plates, or axle guards, of wrought-iron, forming part of the frame of the engine. They form a guide, with the cast-iron slides riveted on to the wrought-iron horn plate, for the axle box to move up and down in, as the springs give way to the weight and jerks of the train. The strain of the engine and carriages comes on the horn plate.

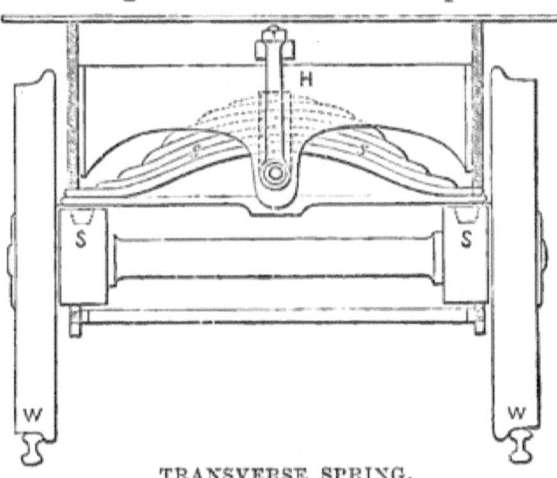

TRANSVERSE SPRING.

This is another method of arranging the spring:—A transverse spring is attached to the framing at H, and carries the weight on its centre. The ends of the spring $s\ s$ rest on the top of the axle boxes at S and S. Their use is to receive the jerks, oscillations, etc., as the engine runs, so that the motion of the engine may be smooth, just as we know, and can feel, the difference between riding in a cart and a carriage, so the springs act in keeping the engine, etc., still.

143. Buffers and Buffer Springs. — Buffers are to receive any sudden shock or strain, so as to give the passengers as little shock as possible.

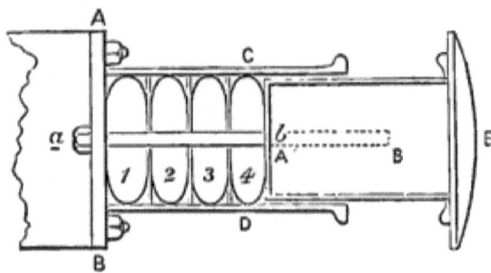

A B is bolted to the buffer beam; within C D are four or more cushions of India-rubber, or India-rubber springs, 1, 2, 3, 4, separated from each other by $\frac{3}{16}$ iron plates, all of which will admit of lateral motion. The bar $a\,b$ passes through all the plates and India-rubber springs. When a shock is received by the buffer E, the springs are compressed and the bar runs up A B, but it is sometimes arranged to drive from right to left. Steel springs are as frequently employed as India-rubber.

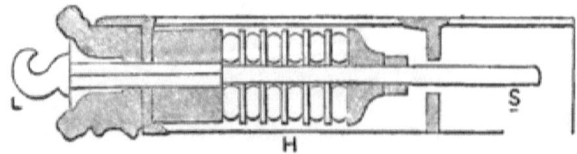

Here we see the arrangement of the *draw bar* and spring for a carriage. At H are the India-rubber springs, L is the hook by which the carriage is attached. The pull of the carriage acts on *s*, and draws the bar towards L, so that the springs are compressed.

For buffing and draw springs, many kinds have been employed. India-rubber springs are formed of circular discs, the buffing and draw rods running through them;

helical and spiral springs, made of steel, and rods acting upon ordinary steel springs, are also used.

144. Draw Bar with Springs.—The draw bar with springs is fitted to engines to receive and take up sudden shocks and strains.

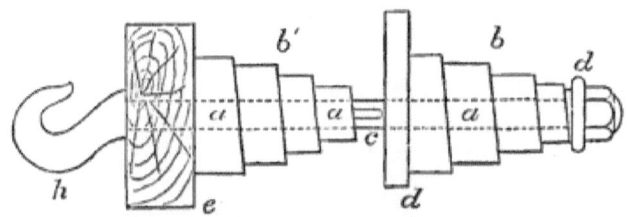

DRAW BAR WITH SPRINGS.

h is a crook or hook to which the carriages are coupled on; $a\ a\ a$ is the *draw bar*, chiefly shown by dotted lines; $b'\ b$ are two steel springs; d' and e are two transverse pieces of the frame, firmly fixed at the same constant distance; e is the buffer beam; c is a cutter to bring up the spring b', while the spring b is brought up by d, a washer close against a nut, as seen in the figure. The action is this: if a pulling strain comes upon the *draw bar*, then the spring b' acts, and is compressed by the cutter c; at the same time the washer d compresses the spring b, thus assisting b' to counteract the strain. The traction spring or draw bar modifies the force of sudden snatches by the engine, which are liable to snap the couplings between the carriages. A plan adopted to resist the strain on carriages, is for the two buffers to act each on the end of an ordinary carriage spring, say from left to right, while the draw bar, to which the carriage is coupled, acts on its centre from right to left.

145. Brakes.—*Brakes* are employed to bring the train to a standstill. They are generally worked by the fireman, although there are *brake vans* with brakes worked by the guards as auxiliaries.

Suppose the handle H pulled to the left (by a screw), then the lever E is drawn towards the left, and with it the lever A B. As A goes to the left, the arm A C jams the brake K against the wheel, while the arm B jams K' against the other wheel, when, friction preventing the revolutions of the wheel, the train is brought to a standstill.

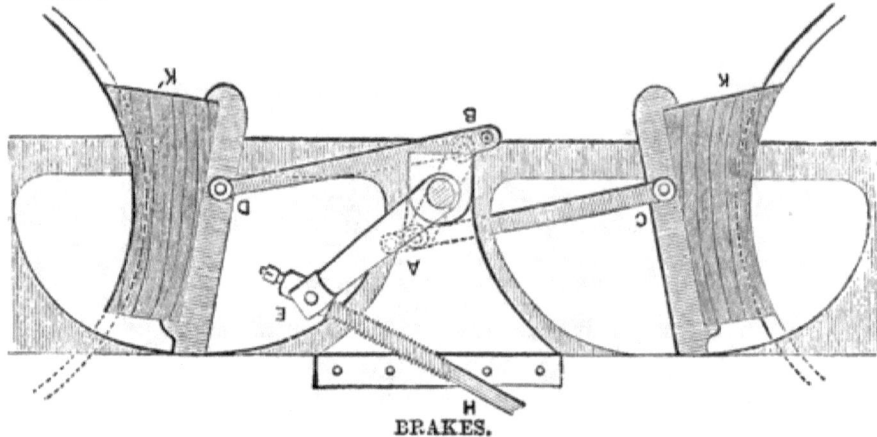

BRAKES.

The brake, which is essentially a screw and lever apparatus, is generally of wrought-iron, except the part which embraces the wheels, which is of wood. Sometimes two or more sledges slide on the rail under the engine. The power developed by the screw and levers is enormous, reaching as high as 500 : 1; so, therefore, if a man turn the screw with the force of half a hundredweight, it acts upon the wheels with a force of more than twelve tons. It does not do to make the leverage excessively great, because the force coming on the frame of the engine, it is liable to be wrenched. The frame must be adapted to bear such extra strains. To save the frame, the force should be thrown on the levers as much as possible, and not on the screw, or the screw should be coarse in its thread, and have a short handle.

CHAPTER IX.

THE ROAD.

Tramway—Railroads—Curves—How the Carriages are Kept on a Curve—Rails—Fish Joints—Gradients—Ballast—Cuttings and Embankments—How Rails are Laid—Two Ways—Broad and Narrow Gauge—To Adapt one Gauge to the other—Switches and Crossings—Tractive Force—Adhesion of Wheels to Rails.

146. The Tram or Tramway is a roadway consisting of long pieces of wood or iron laid down in lines, and prepared to receive the wheels of waggons or trams. They were first used in the North of England and South Wales, for the convenience of carrying coals from the mouth of the pit to seaport and other towns. The way in which they were originally formed may be thus described: First, pieces of oak, 5 or 6 feet long, called *sleepers*, were laid transversely across the track about 2 feet apart; next, longitudinal beams or rails, in lengths of 5 or 6 feet, of sycamore or larch, were laid upon these sleepers, and secured to them by wood pins or trenails; next, these longitudinal pieces of wood were supplanted by rails formed of wrought-iron plates, next cast-iron rails were used. The trams were drawn by horses. Some tramways are constructed of hard stone, as granite and sills with flat iron bars laid upon them for rails. A good idea of one of the earliest form of rails may be obtained by taking a sheet of paper the size of this sheet; first double down one-third of the page longitudinally, turn over the paper and double down the other side in a similar manner; now stand the two pieces perpendicularly to the middle, one will be above and the other beneath. Imagine that the lower one enters the longitudinal rail, and the middle one lies on and is bolted to it, then the wheels of the carriage must be supposed to run on the one standing perpendicularly. Each part was about three inches wide, and of iron one inch or three-quarters thick.

147. **Railroad.** — Railroads are improved tramways. The London and Birmingham Railway is about 30 feet wide on the embankments, and 33 feet in the cuttings; it is wider in the cuttings, because two drains are necessary, one on each side of the line. The average breadth of formation is 18 feet for a single line, and 28 for a double. Space has to be allowed for fencing and ditching. The width on the narrow gauge lines is 4 feet $8\frac{1}{2}$ inches, as North-Western, South-Western, Eastern Counties, etc., and 7 feet on the broad gauge, Great Western. The space between the two lines of rails is 6 feet 6 inches, and is often spoken of as the "six foot way." The sleepers are laid transversely across the road at a distance of from three feet to three feet six inches apart. To the sleepers are fixed the chairs, which are cast-iron supports for the rails. Sleepers are frequently creosoted, or else kyanized, to resist the action of the atmosphere, water, etc. On the broad gauge system, the line is laid with longitudinal sleepers and bridge rails, but the narrow gauge with cross or transverse sleepers and double-headed rails; the rails in the former case being secured directly on to the longitudinal sleepers, whereas those of the latter are supported by cast-iron chairs secured to the cross sleepers. Longitudinal sleepers have been tried on the narrow gauge system, but have not been found to answer so well as the transverse. At least this is the opinion of some experienced engineers.

Railways are *single* or *double*. The double consist of two lines of rails—a *down* line and an *up* line. The down line leads from London, the up line goes to London. To a person looking towards London, the down line is the right hand pair of rails, the up line the left hand pair.

Single lines consist of a single pair of rails used both for the up and down lines. There are double lines at intervals to allow one train to pass another. Lines are constructed on this system for cheapness. The lines should be as level and straight as circumstances will permit.

148. Curves should be of as large a radius as possible; there are but few curves of less than three-eighths of a mile, or 30 chains' radius. The exterior rail of the curve is always elevated—the generic term is *super-elevated*—to counteract the centrifugal force, or otherwise the train might leave the rails. Sharp curves should never be on steep inclines, for the tendency to leave the rails at a curve is as the square of the speed; as a rule, they should be out in the open where they can be well seen, and not in cuttings.

149. How the Carriages are Kept on a Curve.—As an object moves round in a curve, the centrifugal force has a tendency to make it fly off in a straight line. Hence railway carriages, in passing curves, have a tendency to run off the line at the outside. To prevent this, and to keep the flanges of the wheels from the rails, the larger, or outer curve, is raised higher than the inside one, so by this means the carriages are thrown to the opposite side to that on which the centrifugal force would keep them. The *super-elevation* of the outer rail and the conical wheel are thus made to balance the centrifugal force. On the narrow gauge lines, with a wheel three feet in diameter, no super-elevation need be made, unless the curve has a less radius than 1400 feet; on the broad gauge line, with a four feet diametered wheel, the least radius that can be used without super-elevation is double this. The quicker the trains pass a curve the greater must be the elevation of the outer rail.

150. Rails.—The rails are made in many shapes, as seen by the following figures; all these forms are in use, but generally those marked *d* and *c* are preferred. There are many other forms as well as these. At *a* is shown one of the earliest, a plate of iron turned up; the same figure also shows the difference in the arrangement for the running of the wheels on the tramway and railway. At *a* the purpose of one part of the rail is to confine the wheel to the track, and it is evident that much tractive force might be expended in the wheels grating against the rail; but at *g*, the modern arrangement, we see that the wheel is kept on the rail by a flange on the wheel. To the sleepers are fixed the chairs, or chocks, of cast-iron, into which fit the rails, kept in their places by iron spikes. The ends of the rails are secured to each other by a *fish* or *fish plate*, two being used, one on each side, and bolted together by four bolts.

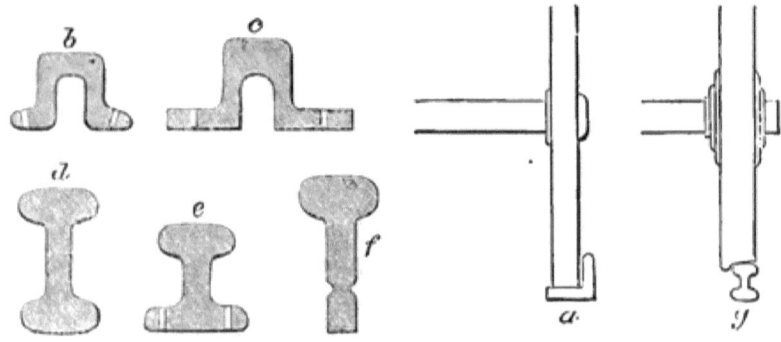

FORM OF RAILS.

151. Jointing of Rails: The Fish Joint.—The two ends of any two adjoining rails are not placed close together, but a small space is left between for expansion. The joint is obviously the weakest part of the rail. The fish joint is intended to give it stability.

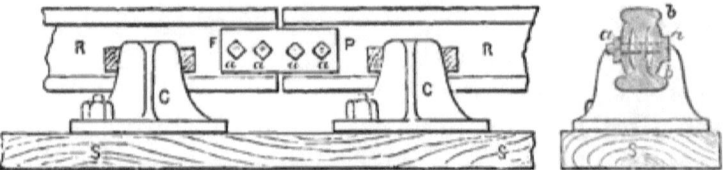

Rails are "fished" by having four holes—*a a a a*—punched in them, and then the fish plate F P is fastened on with four bolts; the holes are larger than the bolts, to allow a slight motion caused by the changes of temperature. The fishes are made of wrought-iron, and bear against the top and bottom of the web of the rail, as seen in the section at b and b. Close to each fish, on either side, are two chairs, C and C, firmly bolted to the sleepers S S. The fish joint is found to answer so well that its use is extending rapidly.

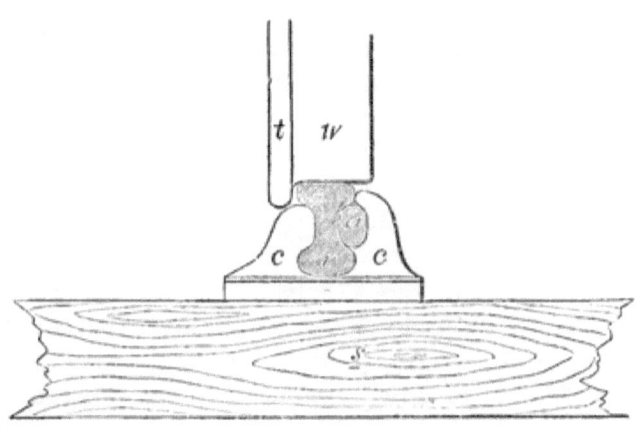

CHAIR, SLEEPER, AND RAIL.

152. Chair, Sleeper, and Rail.—The following simple figure will show how the rail r is fixed in the chair $c\ c$, by means of the wedge a; it also shows the manner in which the flange t of the wheel W clears the chair without touching it, and how it runs on smoothly and evenly without the chair offering any resistance or obstruction.

153. Gradients should not exceed *one* foot rise in a *sixty* feet length, although there are gradients double this, or that rise two feet in sixty. Gradients are very expensive, as extra power, which means fuel, time, and labour, is required to ascend them. When very steep, stationary engines are employed to haul up the trains. Gradients should rise, where practicable, on each side towards a station, for then the weight or gravity of the train will assist the brakes in bringing it to a standstill, while, when leaving, such an arrangement will help to set the train moving. On long inclines there are occasionally level spaces, or benches, to assist the ascending and check the descending train. It is not allowable to place a station on an incline.

154. Ballast.—After the railway is cut, and embankments made, the road is covered with broken hard stones, flint, dry gravel, etc., called ballast, upon this the sleepers are laid. The ballast serves two purposes, it allows all water to drain away, and so the sleepers are kept dry; it also keeps them firm and steady.

155. Cuttings and Embankments.—To save expense, the sides of a cutting should be as steep as possible, for then less earth is moved, but this can scarcely ever be done; no general rule can be given as to what slope should be used, everything depends upon the strata that is being cut through, and not alone upon the top strata, but the bottom strata have frequently to be considered. Most kinds of hard rock will stand vertically, chalk requires a slope of one in three, sand and gravel three feet in two, clay two to one; but there are very great exceptions to every rule. The most troublesome cuttings are where soft clay or wet, soft strata, come under others that are harder and drier, the soft and wet give way, or else the others slip over them, thus giving an enormous amount of trouble, and adding to the expense of the permanent way.

156. How the Rails are Laid.—Two plans, already mentioned, are followed in laying down rails:—(1) That with longitudinal sleepers, which gives a continuous bearing; (2) that with transverse sleepers, in which the sleepers are laid about three feet apart, and the rails supported on chairs.

(1) *The Continuous Bearing.*—Here the rails are firmly secured to long baulks of timber laid in parallel lines, each line inclines a little towards the middle. They are kept at the proper distance apart by transverse pieces of timber, the ends of which are let into the baulks, and then secured by angle plates or wrought-iron knee straps. Sometimes these longitudinal sleepers are laid on trans-

verse or cross sleepers, and thus the advantages of both systems are secured.

(2) *The Transverse Sleeper Bearing.*—This is the system that has been most generally adopted. Sleepers of good strong timber, twelve feet long and six or eight inches thick, properly prepared (see page 137), are laid at intervals of about three feet or three feet six; on each sleeper is securely fixed two chairs at the proper distance, in which the rails are firmly fastened, and so kept in their places steadily, and at a continuously equal distance. Formerly, where stone was plentiful, large blocks of stone were used to fix the chairs to, and thus support the rails.

157. Broad and Narrow Gauge.—The *broad gauge* has a distance of *seven feet* between the two rails on which the carriages run, while the narrow gauge rails are 4 feet $8\frac{1}{2}$ inches apart.

158. To Adapt Broad Gauge to Narrow Gauge.—Great interruption and expense are entailed through railways being of a different gauge. Instead of passengers and goods in bulk being conveyed from the starting place to their destination in the same carriage, much trouble and cost are incurred in changing from one line of rails to another. So much is this inconvenience felt, that gradually, on the Great Western and other lines, a third line of rails is being laid down, so that the inner line of rail and the third serve for the narrow gauge carriages.

159. Fell Railway.—The progress of railway locomotion has compelled engineers to turn their attention to steep gradients, and how best to drive an engine and its carriages up and down steep inclines. Practically, we have returned to Blenkinsop's rail rack. The first plan proposed was to have a middle rail up the steep incline and a pair of wheels on vertical axes gripping the rail on each side, and which, by their forcible revolution, would carry up the train where the ordinary driving wheels would slip without effect. Mr. Fell, for the Mount Cenis Railway, patented a locomotive with horizontal cylinders and two pair of coupled gripping wheels driven direct without the intervention of bevel wheels, the connecting rods that turn the gripping wheels working in a horizontal plane. Powerful springs press the gripping wheels against the centre rail.

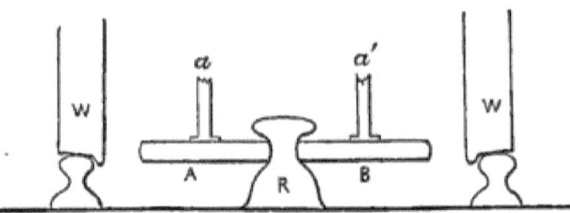

W and W are the wheels of the engine, R the middle rail, A and B the gripping wheels, a and a' their axes.

160. **Switches and Crossings.**—Switches and crossings, or, as they are more commonly termed, *points* and *crossings*, are used for the purpose of allowing the trains to pass or cross over from one line of rails to the other. Several different methods have been devised for doing this. One of the simplest plans, and that most frequently adopted, is to lay down a short line of rails connecting the other two, and thus establishing the desired communication. It is, however, necessary to have ready and expeditious means of connecting and disconnecting this short line with the main line, according as it is intended that the trains shall leave or continue upon the latter; this is effected by the contrivance termed a *switch*, which is shown in our figure.

ab and cd are portions of the rail of the main line, and ef and gh portions of the short line branching from it. All these parts are immovably fixed in the ordinary manner, with the exception of the two rails fi and kl. These, which are termed the tongues of the switch or points, are only fixed at one of their ends f and k, on which they turn as centres; the other ends are tapered away to nearly a point, a slight recess being sometimes cut in the other lines, as at i and l, into which they fit. These tongues are connected together by a bar mno, by means of which they are preserved at such a distance apart, that when either tongue is in contact with the rail near it, the other shall be removed from the one opposite a sufficient space to allow the engine or carriage wheels to pass between. (Suppose the train to come in the direction of the arrow.) In order to keep the train on the main line, or to leave the same and enter the

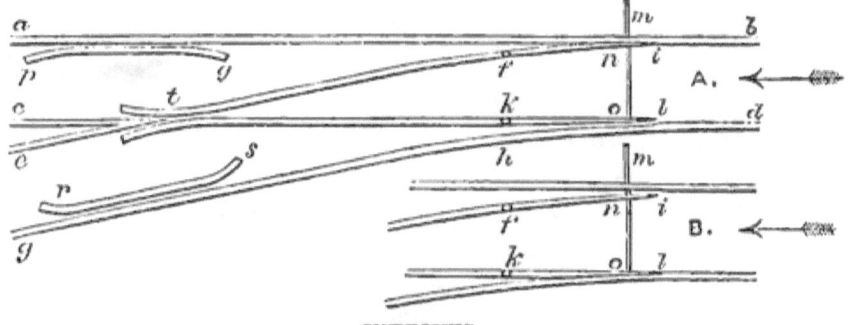

SWITCHES.

branch line, it only becomes necessary to move the bar mno. When mno, or the bar which moves the switch, is in the position as shown at A, the carriages will leave the main line; but if shifted into the position shown at B, then they will continue on their course along the main line. It will assist the student to understand what has been said, if he will consider that the flange of the wheel bears against the inside of the rail. It is usual to have the points so arranged that they are kept in the position shown at B (where the main line is not interrupted) by a self-acting weight, the attendance of a pointsman being necessary to move them into the position A, if it is desirable that the train should go off the main line. Two guard rails, pq and rs, are employed to prevent the flanges of the wheels from striking against the point where the two lines intersect each other at t.

161. Adhesion of Wheels to the Rails.—It was a great difficulty with early locomotive engineers as to how they should secure a proper amount of friction between a smooth wheel and a smooth rail. Hence, in early locomotive engineering, we find geared wheels to the locomotives working in a rack on the tramway. A difficulty did or does exist; for at slow speeds, with full pressure of steam on the piston, it is true that the ordinary adhesion of a single pair of wheels loaded with two or three tons only is nearly useless for any practical purpose. From Mr. G. Rennie's experiments on friction, and the testimony of practical men, it is found that, with extremely light loads upon the driving wheels, there is not sufficient adhesion. Were we now to employ weights of only two or three tons upon the driving wheels of locomotive engines when working at slow speeds, means would have to be provided to prevent slipping. Who has not seen the driving wheel slip when the engine is starting? The adhesion of the wheels to the rails is reckoned at from one-fifth to one-tenth of the load, according as the rails are clean, perfectly wet, perfectly dry, or partly wet.

It has been found that a maximum adhesion upon a clean dry rail of three-tenths, and even three-eighths of the weight on the driving wheels, is occasionally attained. This, of course, is much more than has been counted on by engineers. A better knowledge than was formerly possessed as to the amount of adhesion between the driving wheels and the rails has led to the working of steeper inclines, until, as an extreme case, loads have been taken in practice up *gradients* of one in ten, and no inclination less steep than one in forty is now considered a serious obstacle to the practical working of a large traffic.

162. Tractive Force.—The absolute tractive force required to draw a carriage over a good macadamized road is $\frac{1}{30}$ of the load; but in locomotives, at slow speeds on level rails, it is considered to be about $\frac{1}{300}$ of the load. But, of course, the more rapid the speed the greater the tractive force required. The resistance due to the atmosphere increases very rapidly. It is from 12 to 15 lbs. per ton on a train moving at the rate of 30 miles an hour. At 44 miles per hour, the resistance to train and engine is about 24 lbs. per ton; at 60 miles, 29 lbs. per ton. On rough roads the resistance due to the atmosphere increases as the square of the speed.

The longer the crank of an engine, and the shorter the radius of the driving wheel, the greater the proportion of the pressure on the pistons which will be exerted as tractive force on the rails. The tractive force varies from 6000 lbs., in the case of an express engine, to 20,000 lbs. in the case of a goods engine, and these are not the extremes. Supposing the speed of the pistons to be the same, the express engine would move fastest, because its tractive force is quite sufficient, and the driving wheel much larger in diameter, than that of the goods engine. To exert a great tractive force the driving wheels of an engine must, by their friction upon the rails, have an adhesion equal to the tractive force. For instance, if an engine is to advance, the tractive force being 9 tons, the driving wheels must not slip until the resistance amounts to the same 9 tons. This adhesion is secured by providing sufficient weight upon the driving wheels. On a clean, dry rail as much as one-fourth, and even more, of the total weight on the driving wheel is available for adhesion. One-sixth is, considering the ordinary condition of the weather and other contingencies, quite enough to allow. When half wet, the adhesion is less than when thoroughly wet. They are, in fact, what is termed greasy, and we must not reckon upon more than one-tenth or even less.

CHAPTER X.

INDICATOR AND SLIDE DIAGRAM.

163. The Indicator, an instrument invented by Watt, is used to ascertain the internal condition of the engine, the state of the vacuum, the amount and variations in the pressure of steam at every point of the stroke, the cushioning, the condition of the slides, whether there be too much or too little lap or lead, whether they are leaky or properly set, whether ports are closed and opened at the proper time —in fact, it tells us the power and all the faults by which that power is impaired. It may also be attached to the air pump, the hot well, the condenser, etc., when it will tell us the nature of the pressures there existing. It has been very much modified since the time of Watt, to better adapt it to its purpose. The figure given of it is from one of Richard's indicators, which exhibits the latest improvements.

In its simplest form, the indicator consists of a cylinder with a piston, the top being open to the atmosphere, and a spring to keep the piston down to its work. A diagram is taken on a piece of paper to tell us all we wish to learn. This piece of paper is fastened round a barrel, which moves through nearly a whole revolution and back again as the engine makes one stroke.

In Richard's indicator, A is a screw to fasten the indicator into the cylinder. The handle is to open the connection between the cylinder and the indicator, and thus allow steam to enter B D, the cylinder of the indicator. The piston a and piston-rod b of the indicator are shown by dotted lines. The slanting dotted lines are intended to show the spring which keeps the piston down, and against which the steam has to act in forcing up the piston a. In the actual indicator, the piston is not so simple as shown here, but is conical and truncated; B C is the barrel round which the paper is

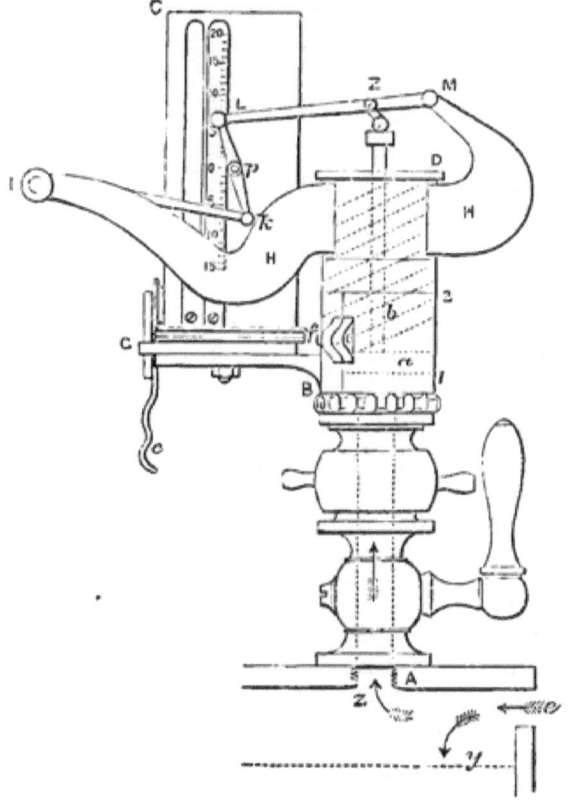

INDICATOR.

wrapped. The graduated scale is to measure the pressure of steam and the vacuum. Within this barrel is a spring, so that when the barrel has moved nearly round once while the piston goes up, the force of the spring causes it to return as the indicator piston goes down. Round the pulley f G passes a string to give motion to the barrel. This string is attached to the crosshead of the cylinder (or the radius bar), and the motion is reduced in its travel to suit the card barrel. While the piston of the indicator moves up only one to two inches, the piston of the cylinder moves several feet. The barrel has to move round four or five inches in the same time. The motion is reduced by levers when taken from the piston crosshead. If the length of the diagram be three inches, and the stroke three feet or thirty-six inches, we have only to proportion the levers as $3:36$ or $1:12$, and the required motion is found. The indicator barrel is moved round by the string (shown in the figure, being attached to its proper relative position on the lever, and) actuating the pulley $f\,G$, and with it the barrel. The arm H H is to carry the parallel motion I k L M, the pencil being at p. The reason of this arrangement (*i.e.*) of having a parallel motion, is that while the stroke of the indicator is (say) only from 1 to 2, the pencil is required to move up and down from the lower fifteen to (say) twenty-five. The head of the indicator piston, being attached to the lever M L at Z, multiplies the motion of the indicator in the proportion of M Z to Z L. In Richard's indicator this multiplier is about three and a half; in fact, this is the essential difference between Richard's and other indicators, such as M'Naught's, Maudslay & Field's, etc., that the motion is magnified, and therefore the pencil more sensibly indicates the least variation of pressure or action.

The action of the indicator must now be traced. Supposing the indicator is attached to the cylinder, but not placed in communication with it by turning the handle, and that the cord c is fastened to the lever at the head of the piston-rod, then it will move the barrel from right to left, and a straight horizontal line will be drawn by the pencil, as A B in next figure—it is generally customary to let the pencil mark this line several times. The line is called the atmospheric line, because it coincides with the atmospheric pressure; all parts of the diagram above that line show pressure above the atmosphere, all parts below it show the vacuum, hence the top part of the diagram is called the "steam" and the bottom the "vacuum." Again: supposing the barrel were still, and the steam admitted to the indicator, the pencil would be driven straight up, or a vertical line would be traced. We see that if the barrel only move, a horizontal line is traced, while if the indicator piston only move, a vertical one is made; therefore, when both move together, we shall have a line compounded of the two motions, and if the one is continually changing, it will not be a diagonal motion.

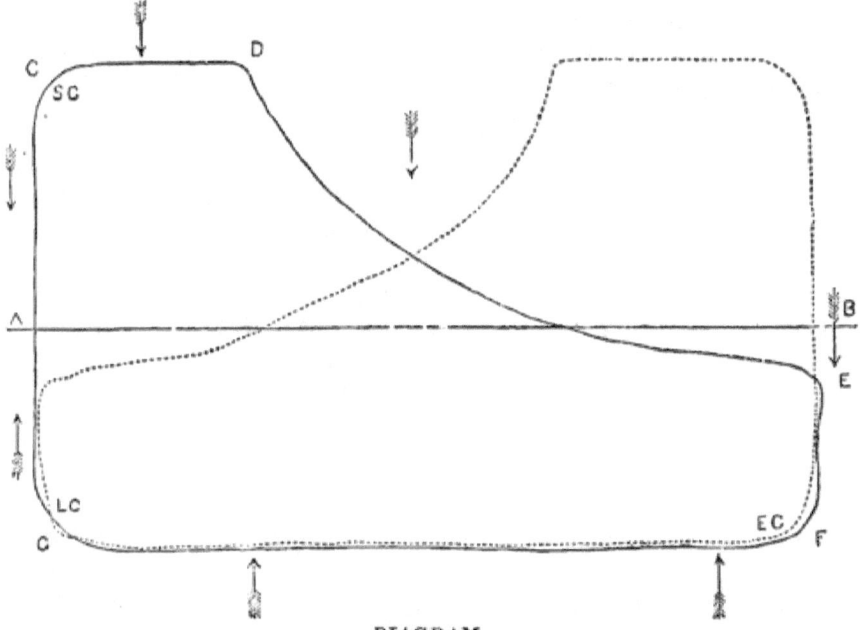

DIAGRAM.

Let us suppose the indicator is attached to the top of the cylinder, and that steam enters the upper port *e* as the piston comes to the top of its stroke. The moment steam enters the cylinder it drives the piston down, but at the same time it enters the indicator, and drives the piston of the indicator up.

Let us suppose the pencil (when air is in both sides of the piston) stands at A on the foregoing figure, then the line A B, which will be traced by the barrel moving nearly the whole way round, is the atmospheric line.

Now let us suppose the top port *e* opened at the instant the top of the indicator is turned, then steam will rush in, in the direction shown by the arrows; in the direction *y* to drive down the piston, and in the direction *p* to drive up the piston of the indicator. Steam coming in instantaneously drives up the pencil, and the line from A to C will be drawn (C is called the starting corner). Now steam continues rushing in at its normal pressure and the piston of the engine goes down, while on the indicator piston the pressure is continuous, so therefore the pencil remains at the same height; and as the barrel moves round, the line from C to D is drawn. When the pencil gets to D, the slide has come down again and closed the port, so that the steam is left to expand; and of course as it expands its pressure decreases, the engine piston continues to go down, and the pressure, becoming less and less in the indicator, the pencil gradually falls lower and lower to E.* When it gets to E, the slide still falling, the upper port *e* is opened to the exhaust, and the steam rushes out in a contrary direction to the arrows, the pencil, therefore, immediately falls to F (the eduction corner). Now there is a vacuum above the piston of the engine, and below that in the indicator, and the engine piston begins to rise up; all the time it is rising, there being no steam or pressure in the indicator (or less than no pressure), the pencil, having fallen to its lowest point, is still, and traces the vacuum line F G to the lead corner G. Against the pencil gets there, the piston has arrived at the top of its stroke, the cushioning then takes place, and the pencil rises at once to A, or else the lead comes into action by the rising of the slide, and drives the indicator piston, and with it the pencil, to A.

The action of the indicator has been traced through an up and down stroke, or a complete revolution of the

* We are supposing a long D slide is used. In reading the paragraph, the student must consider both this and the previous figure.

crank, and we see that the varying pressure in the cylinder is faithfully translated by the indicator, and rendered visible to the eye.

The indicator is absolutely necessary if we are to know the pressure of steam when it is performing its work. The Bourdon gauge or other contrivance, when correctly graduated, will always tell the boiler pressure; but it must be well understood that the boiler pressure seldom or never corresponds to that in the cylinder, it is less. This diagram is supposed to be taken from the top of the cylinder, and the arrows show the direction in which the piston of the engine is moving when that part of the diagram is being traced. The dotted diagram shows one taken from the bottom of the cylinder.

The corners of the diagram are the points to which attention must be directed to find out any defects. In the diagram from a non-condensing engine, the whole of the curve is above the atmospheric line; but in a condensing diagram part is above the atmospheric line and part below.

164. Indicator Diagram of the Locomotive. — The action of the valve in the distribution of steam, as we have already hinted, is regulated by the lap, lead, and travel. When these are given, a diagram will show us at what point of the stroke the steam is admitted, cut off, exhausted, and compressed or shut in. When the link motion is fitted, the steam is cut off earlier by shortening the travel of the slide. This is done in such a manner that, however much the travel of the slide is reduced, the lead is always the same, or at least as at full gear. With the shifting link, it is a little more. When the travel is shortened, not only is the steam cut at an earlier point of the stroke, but it is exhausted earlier, admitted earlier, and the exhaust port is closed earlier during the return stroke. Thus, shortening the travel of the slide causes everything connected with the distribution of steam to be done earlier.

No. 1 was taken with the shifting link in full gear in the first notch of the sector, No. 2 in the second notch, etc.

Taking No. 1 first, we must understand that the port

began to open for the admission of steam at the point A, about $\frac{3}{16}$ of an inch before the beginning of the steam stroke, the line runs up instantaneously to B in time to commence the steam stroke at the full pressure. While the pencil runs from B to C the steam is at a continuous pressure of 38 lbs., as shown by the scale at the side. At C the steam is suppressed or cut off, and while the piston moves the perpendicular distance between C and D

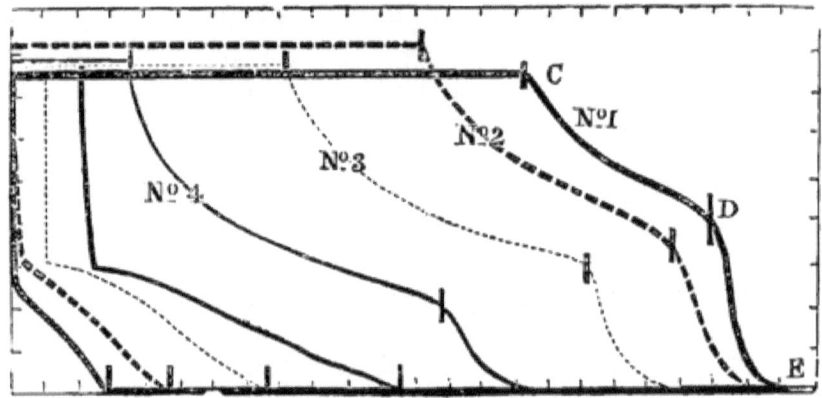

($4\frac{1}{2}$ inches), the enclosed steam expands behind it, rapidly decreasing in pressure, as indicated by the falling line C to D. At D, when the piston has yet to travel the perpendicular distance from D to G, the port is opened to the exhaust, *i.e.*, it is opened when the piston has yet to travel *three* inches, the pressure therefore quickly decreases, as shown by the falling line from D to E. During the return stroke, the steam continues to exhaust into the atmosphere, and the atmospheric line E F is traced; but ordinarily the diagram seldom coincides with the line, as we have it here, for there is a certain amount of back pressure. When the piston gets to F, within three inches of the end of the return stroke, the exhaust port is closed, and the piston continuing its motion, the cushioning takes place, and the pressure of the pent-up steam increases, as shown by the rising curve F to A; when at A the steam is re-admitted, and the curve traced again.

MISCELLANEOUS EXERCISES.

1. Convert 40° Fahrenheit to Centigrade, and 50°C. to F.

$$\begin{array}{ll} 40°\text{F.} & 50°\text{C.} \\ \underline{32} & \underline{9} \\ 8 & 5)\overline{450} \\ \underline{5} & \overline{90} \\ 9)\overline{40} & \underline{32} \\ \overline{4\tfrac{4}{9}°\text{C.}} & \overline{122°\text{F.}} \end{array}$$

2. Convert 90°C. to F., −10°F. to C., −10°C. to F., and 150°F. to C.   *Ans.* 194°F., −23⅓°C., 14°F., 65⅝°C.

3. A piece of metal expands $\frac{1}{1000}$ part of an inch on being heated one degree; find the length of a rod of iron 20 yards long after being heated from 40° to 300°.

Increase in inches $= \dfrac{20 \times 3 \times (300 - 40)}{1000} = 15\tfrac{3}{5}$ inches.

∴ Total length $= 60$ ft. $+ 1$ ft. $3\tfrac{3}{5}$ in. $= 61$ ft. $3\tfrac{3}{5}$ in.

4. The co-efficient of brass is ·001984, find the increase of length of a bar raised from 28° to 358°.   *Ans.* 7·5 inches.

5. A weight of 3 cwt. is lifted 12 fathoms, how many units of work were done?

Units of work $= 3 \times 112 \times 12 \times 6 = 24464$.

6. The beam of a steam engine weighs 4 tons 12 cwt. 3 qrs. 7 lbs.; it is lifted into its place 42 feet high: what number of units of work were performed?   *Ans.* 436590.

7. A horse can do 33,000 units of work per minute; how many horse-power were required to lift the beam, if it were done in 5 minutes?

Units of work done in one minute $= \dfrac{436590}{5} = 87318$

∴ Horse-power required $= \dfrac{87318}{33000} = 2·646$, *Ans.*

8. Two tons of water are lifted every 3 minutes from the bottom of a mine 275 feet; what is the horse-power of the engine to do the work?   *Ans.* 12⅘ H.-P.

9. Suppose in the last case that the modulus of the engine, or the "relation of the work done to the work applied," is ·6, find the horse-power.   *Ans.* 7$\tfrac{7}{15}$ H.-P.

10. A locomotive engine, weighing 15 tons, is lifted by a crane 21 feet high in 4 minutes; find the number of units of work done by the crane per minute.   *Ans.* 176400.

STEAM.

11. I heat 40 cubic feet of air from 30°C. to 50°; what is the increase of volume, and what is the present volume?
Ans. 42·92 cubic feet.

12. If 900 feet of air be heated through 75° of heat, what is the increase of volume? *Ans.* 247·2 cubic ft.

13. If 500 feet of gas have their temperature lowered through 60°, what is the volume remaining? *Ans.* 409·9 cubic ft.

14. Suppose the boiling point of water on the summit of Mont Blanc is 85°·14C., what is the height of the mountain? The boiling point of water decreases 1°C. for every 1062 feet perpendicular height. *Ans.* 15781 ft.

15. Twelve ounces of iron at 600°C. are placed in 8 oz. of water at 50°C.; how much water is converted into steam, supposing no heat lost in the process? *Ans.* ·526 oz.

16. If 16 oz. of iron at 500°C. are placed in 10 oz. of water at 60°C., and the specific heat of iron is considered as ·1138; how much water will be converted into steam? *Ans.* 611 oz.

17. What is meant by the grate surface of a boiler? If 1 square foot be allowed for each horse-power, how much will be necessary for boilers to supply a pair of cylinders each of 73 inches diameter, the piston moving at the rate of 240 feet a minute (1866)? Find the nominal horse-power.
Ans. 426·32.

18. Give a description of the apparatus by which a boiler is prevented from bursting and collapsing. How is the pressure of the steam in the boiler ascertained? Can the same dependence be placed on an old gauge as on a new one (1865)?

19. What precautions should be taken to prevent boiler flues from collapsing? Give an idea of the pressure they have to sustain, and how should their thickness vary with their diameter?

20. Forty cubic feet of water loses 10° of heat; how much air will this heat 20°? *Ans.* 64972·6 cubic feet.

21. Steam at 24 lbs. pressure is admitted into a cylinder above the piston 50 inches in diameter; find the total pressure on the piston (1) when there is a vacuum below; (2) when the air is freely admitted below. *Ans.* 47124 lbs.: 17671·5.

22. Steam of 30 lbs. pressure is admitted (1) to one side; (2) to the other, the diameter is 45 inches, and diameter of piston-rod 5 inches; find the difference between the pressures on the upper and lower side of the piston. *Ans.* 589 lbs.

23. The steam pressure is 40 lbs. per circular inch, what is the pressure in tons on a 36-inch cylinder?
Ans. 23¼ tons.

24. Describe the feed pump and valves necessary for supplying the boiler of a locomotive. What is the principle of Giffard's injector (1869)?

25. How is a locomotive engine reversed by the use of a double eccentric and link motion? What is the object of the sector with notches cut in it, whereby the starting lever can be held in intermediate positions (1869)?

26. Describe the general construction of a locomotive boiler. Why is the fire box made of copper? How is it attached to the iron shell which surrounds it? How is the roof of the fire box strengthened (1871)?

27. The stroke of the piston of an engine is 24 inches, and the diameter of the driving wheel is 8 feet, what is the mean velocity of the piston when the engine is running at 40 miles an hour (1871)?

At each revolution the wheel goes $8 \times 3\cdot 1416$ feet, 40 miles per hour is $40 \times 1760 \times 3$ feet.

$\therefore$ the train moves in feet per min. $\dfrac{40 \times 176 \times 3}{60}$

$\therefore$ number of revolutions of wheel per min. $= \dfrac{40 \times 1760 \times 3}{60 \times 8 \times 3\cdot 1416}$

But in each revolution of the wheel the piston moves $2 \times 2 = 4$.

$\therefore$ speed of piston $= \dfrac{40 \times 1760 \times 3 + 4}{60 \times 8 \times 3\cdot 1416}$

$= 560$ feet per minute.

28. How does a railroad differ from a tramroad? Describe the method of supporting rails upon cross sleepers, and of joining them securely (1871).

29. State what you know in respect of the arrangement and construction of springs for the three different purposes for which they are fitted to a passenger carriage,—viz., as buffer, draw, and bearing springs (1871).

30. Describe some form of regulation valve for admitting steam into the pipe leading into the cylinders. Where is this valve placed (1871)?

31. Describe the safety valve of a locomotive boiler. Explain Bourdon's gauge for ascertaining the exact pressure of the steam in a boiler (1869).

32. Describe generally the construction of a railroad. How are the tires of the wheels of the carriages shaped? and for what reason? Describe the fish joint (1869).

33. Describe the construction of a locomotive boiler. How is the fire box attached to the barrel of the boiler? In what way is the draught obtained? In a locomotive boiler there are 156 tubes, each 2 inches in diameter and 127 inches long, what amount of heating surface do they give (1870)? *Ans.* $864\cdot 4636$ sq. ft.

34. Show how you would allow for the weight of the lever in adjusting the weight of the safety valve.

35. Explain the importance of balancing the cranks in a locomotive engine. The leading wheel of an engine is $3\tfrac{1}{2}$ feet in diameter, what would be the pull on the centre of the wheel caused by an unbalanced weight of 9 lbs. upon the rim, when the engine was running at 20 miles an hour (1870)?

$$\text{Centrifugal force} = \left(\frac{v}{4\cdot 11 \over D}\right)^2 \times 9 = 137\cdot 5 \text{ lbs.}$$

v = velocity per second.

36. Supposing a train of 60 tons is drawn up an incline of 1 in 100, and the friction is 8 lbs. per ton, find the work due to gravity, friction, and the total power required to draw the train up the incline.

If it rises 1 in 100, the force due to gravity $= \tfrac{1}{100}$.

∴ Force due to gravity on a 60 ton train $= \tfrac{60}{100} = \tfrac{3}{5}$ tons
$= 1344$ lbs.

Also as friction is 8 lbs. per ton
force due to friction $= 60 \times 8 = 480$ lbs.
∴ Force to draw it up the incline $= 1344 + 480$
$= 1824$ lbs.

37. Find the horse-power of a locomotive engine, which, running 40 miles per hour on a level tract, draws a train weighing 70 tons, taking the friction at 8 lbs. per ton, and neglecting the resistance of the air.

$$\text{Distance train moves per minute} = \frac{40 \times 5280}{60} = 3520 \text{ feet.}$$

Resistance due to friction $70 \times 8 = 560$ lbs.
∴ Work of friction per minute $= 3520 \times 560$
This must equal the horse-power in units of work.
∴ Horse-power $\times\ 33000 = 3520 \times 560$
∴ Horse-power $= \dfrac{3510 \times 560}{33000} = 59\cdot 56.$

The students are referred to the work on "Steam and the Steam Engine," in the Advanced Course of this Series, for Examples and Exercises, and the proper method of making the various calculations.

INDEX.

Absorption, 41.
Adhesion of Wheels, 145.
Advantages of Superheated Steam, 36.
Analysis of Sea Water, 33.
Angular Advance, 74.
Ash Pan, 103.
Atomic Force, 15.
Axle Boxes, 131.

Ballast, 142.
Barometer Gauge, 84.
Beam Engine, 66.
Blast Pipe, 104.
Blow-off Cock, 123.
Blow-through Valve, 83.
Bogies, 97.
Boiler, 97.
Boiler of Rocket, 93.
Bourdon's Gauge, 112.
Brakes, 135.
Broad Gauge, 143.
Brining Boiler, 34. [134.
Buffers & Buffer Springs,

Calorimeter, 20.
Capacity for Heat, 19-22.
Cataract, 60.
Centigrade Thermometer Chair, 141. [17.
Chemical Affinity, 16.
Chimney, 107.
Clearance, 55, 99.
Co-efficient of Expansion,
Conduction, 25. [11.
„ of Water, 23.
Condenser and Air Pump, 50.
Condensation, Water required for, 38, 81.
Connecting Rod & Crank 69, 127.
Cornish Valve, 87.

Contraction by Cold,
Convection, 22. [12, 14.
Conversion of Heat into Work, 22. [130.
Counterweights to Wheels
Coupling Rods, 128.
Crampton's Engine, 96.
Crane, Water, 114.
Crank, 69.
Curves, 139.
Cushioning, 55.
Cuttings, 142.
Cylinder, 125.
„ and Crank, 70.
Cylindrical Slides, 72.

Defects of Newcomen's Engine, 49.
Description of Locomotive, 94.
Diagram Indicator, 151-3
Discoveries of Watt, 49.
Dome, Steam, 107.
Double Acting Engine, 54.
Double Beat Valve, 86.
Drain Cocks, 126.
Draw Bar, 135.
Driving Wheel, 129.

Eccentric, 62.
„ Single, 76.
„ Double, 76.
Embankments, 142.
Engine, Beam, 66.
„ Double Acting, 54, 66.
„ High and Low, Pressure, 66.
„ How Worked, 70
„ Newcomen's, 46
„ Savarys, 45.
„ Single Acting, 53

Engine, Starting, 56.
„ To Reverse, 76.
Equilibrium Valve, 87.
Expansion Gear, 79.
„ to carry out, 80.
Escape Valve, 87.

Feed Pump, 68, 119.
Fell Railway, 143.
Fire Bars, 101.
Fire Box, 106.
Fish Joint, 140.
Fly Wheel, 85.
Furnace, 100.
„ Stays, 101.

Galvanic Action, 43.
Gaseous State of Matter,
Gauge, 83, 84, 143. [11.
„ Cocks, 122.
„ Bourdon's, 112.
Glands, 56.
Glass Water Gauge, 122.
Gooch's Safety Valve,
Governor, 58. [106.
„ Marine, 61.
Gradients, 141.
Grease Cocks, 126.
Gridiron Valve, 73.

Heat, 10.
„ Capacity for, 19.
„ Consumption of in liquefaction, 25.
„ Conversion into Work, 26.
„ Expansion by, 12.
„ Latent, 10.
„ Unit, 19.
High Pressure Steam, 95.

Indicator, 148.
Jointing of Rails, 140,

INDEX.

Lap and Lead, 74.
Latent Heat, 10.
  ,,   of Water, 23.
  ,,   of Steam, 24.
Laying of Rails, 142.
Lead, 75.
Linear Advance, 74. [11.
Liquid State of Matter,
Locomotive Boiler, 97.
  ,,   Slide, 71.
Long D Slide, 71.

Man Hole, 109.
Marine Governor, 61.
Mercurial Gauge, 83.
Molecular Force, 15.

Narrow Gauge, 143.
Newcomen's Engine, 46.

Oxidation of Metals, 44.

Parallel Motion, 57.
Piston, 55, 126.
Point of Saturation, 33.
Power of Expansion and Contraction, 14.
Pressure and Temperature, 32.
Pressure of Steam, 26.
Priming, 107.
Pyrometer, 8.

Questions, 27, 36, 64, 88, 155.

Radiation, 41. [41.
Radiators, Good and Bad,
Railroads, 138.
Rails, 139.
Ramsbottom's Self-Filling Tender, 117.

Reaumer's Thermometer, 17.
Reciprocity of Radiation and Absorption, 42.
Regulator, Steam, 108.

Safety Valve, 105. [111.
Salter's Spring Balance,
Salt in Sea Water, 33.
Sand Cock, 130.
Saturation, 33.
Savary's Engine, 45.
Scale, to Prevent, 35.
Screw Plugs, 123.
Scum Cocks, 123.
Sector, 113.
Self-filling Tender, 117.
Smoke Box, 103.
Snifting Valve, 54.
Specific Heat, 22. [38.
  ,,   Gravity of Steam,
Staying Furnace, 101.
Sleepers, 141.
Slides, 70.
  ,,   Cylindrical, 73.
  ,,   Gridiron, 73. [75.
  ,,   Locomotive, 71,
  ,,   Long D, 71.
  ,,   Seawards, 72.
  ,,   Short D, 72.
Springs, 131.
Steam Cut Off, 74.
  ,,   Definition, 9.
  ,,   Dome, 107.
  ,,   Elastic, 10.
  ,,   Expansive Working, 51.
  ,,   Full, 74.
  ,,   High Pressure, 35.
  ,,   Invisible, 9.
  ,,   In Contact with Water, 37.

Steam, Latent Heat, 24.
  ,,   Measure of Pressure, 36.
  ,,   Regulator, 108.
  ,,   Specific Gravity, 38.
  ,,   Superheated, 36.
  ,,   and Vapour, 31.
  ,,   Whistle, 110.
Stephenson's Driving Gear, 91. [76.
Stephenson's Link Motion,
  ,,   Rocket, 90.
Stuffing Box and Glands, 56. [144.
Switches and Crossings,

Temperature of Bodies, 16.
Thermometer, 16.
Throttle Valve, 59.
Through Tie Rods, 98.
Tire of Wheel, 129.
Tramway, 137.
Tractive Force, 146.
Tubes, 98.

Unit of Heat, 19.

Vacuum Gauge, 113
Vapour and Steam, 31.

Water for Condensation, 38.
Water for Cocks, 126.
  ,,   Crane, 114.
  ,,   Tank, 114.
Watt's Discoveries, 49.
  ,,   Single Acting Engine, 53.
Wheels, Adhesion of, 145.
Whistle, Steam, 110.

www.ingramcontent.com/pod-product-compliance
Lightning Source LLC
Chambersburg PA
CBHW030251170426
43202CB00009B/709